Springer Theses

Recognizing Outstanding Ph.D. Research

Aims and Scope

The series "Springer Theses" brings together a selection of the very best Ph.D. theses from around the world and across the physical sciences. Nominated and endorsed by two recognized specialists, each published volume has been selected for its scientific excellence and the high impact of its contents for the pertinent field of research. For greater accessibility to non-specialists, the published versions include an extended introduction, as well as a foreword by the student's supervisor explaining the special relevance of the work for the field. As a whole, the series will provide a valuable resource both for newcomers to the research fields described, and for other scientists seeking detailed background information on special questions. Finally, it provides an accredited documentation of the valuable contributions made by today's younger generation of scientists.

Theses are accepted into the series by invited nomination only and must fulfill all of the following criteria

- They must be written in good English.
- The topic should fall within the confines of Chemistry, Physics, Earth Sciences, Engineering and related interdisciplinary fields such as Materials, Nanoscience, Chemical Engineering, Complex Systems and Biophysics.
- The work reported in the thesis must represent a significant scientific advance.
- If the thesis includes previously published material, permission to reproduce this must be gained from the respective copyright holder.
- They must have been examined and passed during the 12 months prior to nomination.
- Each thesis should include a foreword by the supervisor outlining the significance of its content.
- The theses should have a clearly defined structure including an introduction accessible to scientists not expert in that particular field.

More information about this series at http://www.springer.com/series/8790

Sara Correia Carreira

Rapid Cell Magnetisation Using Cationised Magnetoferritin

Doctoral Thesis accepted by
the University of Bristol, UK

Springer

Author
Dr. Sara Correia Carreira
School of Oral and Dental Sciences
University of Bristol
Bristol
UK

Supervisor
Dr. Annela Seddon
HH Wills Physics Laboratory
University of Bristol
Bristol
UK

ISSN 2190-5053 ISSN 2190-5061 (electronic)
Springer Theses
ISBN 978-3-319-60332-2 ISBN 978-3-319-60333-9 (eBook)
DOI 10.1007/978-3-319-60333-9

Library of Congress Control Number: 2017943101

Printed on acid-free paper

This Springer imprint is published by Springer Nature
The registered company is Springer International Publishing AG
The registered company address is: Gewerbestrasse 11, 6330 Cham, Switzerland

Supervisor's Foreword

I am delighted that Springer has chosen to recognise Sara Correia Carreira's work by publishing it as part of the Springer Theses series. Sara joined the Bristol Centre for Functional Nanomaterials (BCFN), where I am the Director, in 2011, and chose to undertake her Ph.D. research project in my research group, and that of Prof. Walther Schwarzacher, also within the School of Physics. As is so often the case, our initial project proposal, of functionalisation of magnetoferritin with lipid molecules, turned out to be less straightforward than anticipated, and so a new direction was sought. The result of this is the work presented in this book, which represents a true interdisciplinary project, and for which Sara was very much the driving force.

The ability to label cells magnetically, either for applications in imaging, or for manipulation and separation is a rich vein of interdisciplinary research, with potential impacts in MRI scanning, regenerative medicine, diagnostics and pathogen detection. However, often the act of loading superparamagnetic iron oxide nanoparticles (SPIONS) onto cells in sufficient quantity has adverse effects on cell survival, and in many cases to achieve sufficient concentrations to be of use can take more than 24 h to achieve. Sara had studied magnetoferritin, a SPION which could be nucleated within the core of the naturally occurring iron storage protein, ferritin, as a possible route to developing biocompatible monodisperse magnetic particles. She realised that the ability to alter the surface chemistry of the protein cage within which the SPION sat may be the key to developing a rapid and non-cytotoxic label which could herald a significant improvement over current labelling strategies. She achieved this by cationising the surface of the protein, vastly increasing its positive charge which had the effect of making this new protein SPION attach strongly to membrane surfaces through electrostatic interactions. The result of this meant that cells could be labelled effectively at low doses, so that cell viability was not compromised, and labelling times were reduced from a number of days down to less than one minute.

Armed with this knowledge, Sara set out to show how this labelling strategy might be applied. She chose human mesenchymal stem cells as her initial target and was able to show that the binding of her protein SPION not only effectively labelled the cells but that this magnetisation persisted for a period of weeks. In addition, MRI studies showed significantly enhanced T2 contrast enhancement. Most critically, the cells could be differentiated into different tissue types, including cartilage, which is something that had not been demonstrated with previous SPION labelling techniques.

Sara then showed that this labelling method could have applications in the capture and concentration of bacteria from dilute samples—an issue in many diagnostic devices where a very small population of bacteria of interest may be present in a large sample volume. She showed that common bacteria, including *E. coli* and *S. aureus*, could be labelled with her SPION and captured in a magnetic column with 99.97% efficiency and that *E. coli* could be concentrated from extremely dilute solutions and cultured. Each of these steps which Sara was able to elucidate has an impact on our ability to develop diagnostic tools which may help in the fight to overcome antibiotic resistance.

The methodology which Sara has developed shows great versatility and flexibility in its applications. The non-specific nature of the electrostatic binding means that it is (as she has shown) applicable to a number of cell types, whilst the ability to manipulate the surface chemistry affords the opportunity to introduce more specific functionalisation if required. Additionally, the ferritin protein cage within which the SPION resides is not limited to an iron nanoparticle—it is relatively facile to envisage this cage as a vector which can be loaded with a number of different functional molecules, opening up applications far broader than could possibly be described in one thesis.

What has made Sara stand out for me as a student is more than just her technical ability. At BCFN, we have been actively trying to grow a culture of true interdisciplinarity, with students having a wide network of academics and their peers to draw upon for inspiration; Sara embodies this through her proactivity and the interactions she has initiated with students from other BCFN cohorts, and our wider network which have enabled her to complete this work. We are particularly indebted to BCFN alumnus Dr. James Armstrong and his supervisor, Dr. Adam Perriman, whom Sara worked with on the stem cell project, and Dr. Jim Spencer, with whom she undertook the bacteria work. On a personal note, Sara was my first Ph.D. student and has set a very high benchmark for those who will come after her! The BCFN have been very fortunate to have had the chance to train a student of her calibre, and this is reflected in the work presented in this volume.

Bristol, UK
May 2017

Dr. Annela Seddon

Abstract

Magnetic cell labelling with superparamagnetic iron oxide nanoparticles (SPIONs) facilitates many important biotechnological applications, such as cell imaging and remote manipulation. However, to achieve adequate cellular loading of SPIONs, long incubation times (24 h and more) or high exposure concentrations are often employed, which can adversely affect cell function. This work aimed at developing a facile surface functionalisation strategy that enables rapid and versatile cell magnetisation using low exposure concentrations. It was found that chemical cationisation of magnetoferritin produced a novel, highly membrane-active SPION that magnetised human mesenchymal stem cells (hMSCs) and two bacterial species using incubation times as short as one minute.

In hMSCs, magnetisation persisted for several weeks in culture and provided significant T2 contrast enhancement during magnetic resonance imaging. Exposure to cationised magnetoferritin did not adversely affect the viability, membrane integrity, proliferation and multi-lineage differentiation capacity of hMSCs. Significantly, chondrogenesis was not inhibited, which is a differentiation pathway that is often affected by SPION exposure.

A one-minute incubation with cationised magnetoferritin also magnetised the gram-negative bacterium *Escherichia coli* and the gram-positive *Staphylococcus aureus*, such that these bacteria could be captured in a magnetic column with an efficiency of at least 99.97%. Importantly, *Escherichia coli* could be concentrated from a very dilute suspension ($<10^2$ cfu mL^{-1}), which is an important pre-processing step for rapid, microfluidics-based pathogen detection.

The combination of synthetic ease and flexibility, the rapidity of labelling and absence of cytotoxicity make this novel SPION system an easily accessible and versatile platform for a range of cell-based therapies in regenerative medicine, as well as an attractive tool for rapid pathogen isolation.

Acknowledgements

I would like to thank Dr. Annela Seddon and Prof. Walther Schwarzacher for providing excellent supervision and intellectual stimulation. Their positive attitude, encouragement and support were very much appreciated, particularly in those hard times of repeated failure when everything seemed to go wrong and I was doubting myself. I would also like to thank Dr. Adam Perriman and Dr. James Armstrong for welcoming me into their laboratories and inspiring many brilliant ideas—if only we had time to pursue them all! It has been such a pleasure working in an environment where ideas and resources are shared so openly. I hope to always encounter this collaborative spirit, which I believe enables the greatest of our scientific endeavours.

A big "thank you" to everyone in the Bristol Centre for Functional Nanomaterials and in the Micro- and Nanostructures Group in the School of Physics. Fun times were had over cake, Spanish goodies served on paper plates, treasure hunts, and many a barbecue. A particular "thank you" goes to my dear friend and former colleague Janina Moereke, who made me feel welcome when I first joined the Micro- and Nanostructures Group and shared many proud moments with me. I would also like to thank all the other great people that I had the privilege to meet in the School of Physics, such as Dr. Carsten Putzke, Dr. Stuart Bellamy, Dr. Ruth Oulton and Sebastian Knauer. Their friendship and support has been simply wonderful.

Finally, my greatest and most heartfelt "thank you" goes to my dad, my mum and my brother. My dad's curiosity encouraged me, my mum's strength empowered me and my brother's ingenuity inspired me. I don't think I would have become a scientist without them. And that would have been a shame, 'cause let's face it: this is the best job ever!

Contents

Abbreviations

Listed below are the abbreviations used in this thesis. They are generally defined where they first appear in the text. Commonly used units, chemical symbols, company names and brands have been excluded.

ABC	Avidin and biotinylated horseradish peroxidase macromolecular complex
ALP	Alkaline phosphatase
ANOVA	Analysis of variance
ApoF	Apoferritin
ATP	Adenosine triphosphate
BSA	Bovine serum albumin
cat-ApoF	Cationised apoferritin
cat-MF	Cationised magnetoferritin
CB11B	Peptide with the amino acid sequence: KVGPSGA[Hyp]GEDGR [Hyp]GP[Hyp]GP
cfu	Colony-forming units
cgs	Centimetre-gram-second system of units
COL2-3/4	Monoclonal mouse primary antibody against denatured type II collagen
DAB	3,3-diaminobenzidine
DLS	Dynamic light scattering
DMEM	Dulbecco's modified eagle medium
DMMB	1,9-dimethylmethylene blue
DMPA	*N*,*N*′-dimethyl-1,3-propanediamine
DMSO	Dimethyl sulfoxide
DNA	Deoxyribonucleic acid
DPX	Di-*n*-butyl phthalate in xylene
ECM	Extracellular matrix
EDC	*N*-(3-dimethylaminopropyl)-*N*′ ethylcarbodiimide
EDTA	Ethylenediaminetetraacetic acid

ELISA	Enzyme-linked immunosorbent assay
EPC	Endothelial progenitor cell
ETR	Electron transfer reagent
FBS	Foetal bovine serum
GAG	Glycosaminoglycan
H&E	Haematoxylin and eosin
HEPES	4-(2-hydroxyethyl)-1-piperazineethanesulfonic acid
HIV	Human immunodeficiency virus
hMSC	Human mesenchymal stem cell
HPC	Haematopoietic progenitor cell
ICP-OES	Inductively coupled plasma optical emission spectroscopy
IEC	Ion exchange chromatography
IgG	Immunoglobulin G
INT	2-(4-iodophenyl)-3-(4-nitrophenyl)-5-phenyl-2*H*-tetrazolium
IR	Inversion recovery
ITS	Insulin transferrin selenium
LB	Lysogeny broth
LDH	Lactate dehydrogenase
LPS	Lipopolysaccharide
m/z	Mass-to-charge ratio
MACS	Magnetic cell separation
MALDI	Matrix-assisted laser desorption ionisation
MES	2-(*N*-morpholino)ethanesulfonic acid
MF	Magnetoferritin
MPMS	Magnetic Property Measurement System
MR	Magnetic resonance
MRI	Magnetic resonance imaging
MSC	Mesenchymal stem cell
MTS	3-(4,5-dimethylthiazol-2-yl)-5-(3-carboxymethoxyphenyl)-2-(4-sulfophenyl)-2H-tetrazolium
NAD	Nicotinamide adenine dinucleotide
NAD+	Oxidised form of NAD
NADH	Reduced form of NAD
NPC	Neural progenitor cells
OPC	Oligodendrocyte progenitor cell
PBS	Phosphate-buffered saline
PCR	Polymerase chain reaction
PDB	Protein Data Bank
PDI	Polydispersity index
PEG	Polyethylene glycol
PG	Peptidoglycan
PGA	Polyglycolic acid
PLL	Poly-(L-lysine)
RES	Reticuloendothelial system
RF	Radio frequency

RGD-4C	Peptide with the amino acid sequence; ACDCRGDCFCG
ROI	Region of interest
ROS	Reactive oxygen species
SE	Spin echo
SEC	Size exclusion chromatography
SI	Système International d'Unités (French for international system of units)
SPION	Superparamagnetic iron oxide nanoparticle
SQUID	Superconducting quantum interference device
TAT	Peptide with the amino acid sequence: GRKKRRQRRRPQ
TE	Echo time
TEM	Transmission electron microscopy
TGF β3	Transforming growth factor β3
TI	Inversion time
TOF	Time-of-flight
TPCK	Tosyl phenylalanyl chloromethyl ketone
TR	Repetition time
Tris	Tris(hydroxymethyl)aminomethane
USPION	Ultrasmall superparamagnetic iron oxide nanoparticle
UV	Ultraviolet
α-MEM	Minimum essential medium eagle α-modification

Chapter 1
Introduction

In this chapter the concepts that form the basis of this work will be introduced, as well as the motivations for and the structure of the thesis. First, the physical characteristics of magnetic nanoparticles will be discussed, alongside biomedical applications of these materials. Then, the biology of the two cell types, which were used in this work, is introduced (mesenchymal stem cells and bacteria). This is followed by a discussion of motivations and strategies for cell labelling with magnetic nanoparticles. After that, an introduction to magnetoferritin is given, as well as the chemical basis of protein cationisation, which is an important method employed in this work. Finally, the motivations underlying the thesis are summarised and an overview is given over the hypotheses tested.

1.1 Magnetism

Several good textbooks on the physical principles of magnetism and magnetic materials are available, e.g. the works of Spaldin [1] and Krishnan [2] give an excellent introduction to fundamentals and applications.

1.1.1 Origins of Magnetism

The magnetic properties of materials are a consequence of magnetic moments associated with individual electrons. These magnetic moments arise from the orbital motion of electrons around the nucleus and the electron spin around its own axis (Fig. 1.1). Thus, each electron in an atom is essentially a small magnet with a permanent orbital and spin magnetic moment.

The fundamental magnetic moment is the **Bohr magneton** μ_B which is of magnitude 9.27×10^{-24} Am^2. For each electron in an atom the spin magnetic

S. Correia Carreira, *Rapid Cell Magnetisation Using Cationised Magnetoferritin*,
Springer Theses, DOI 10.1007/978-3-319-60333-9_1

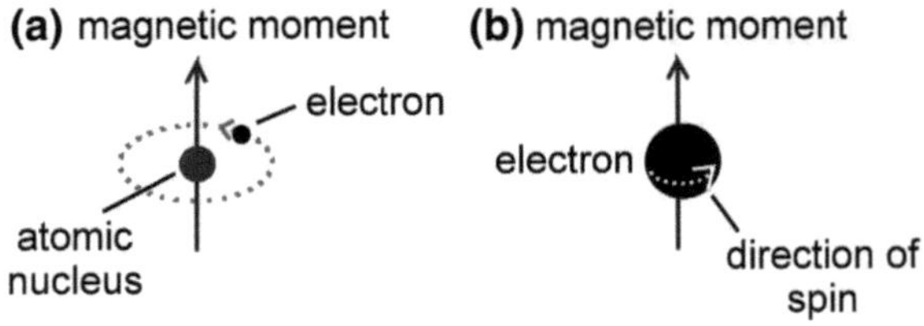

Fig. 1.1 Origin of the atomic magnetic moment. Magnetic moment associated with (**a**) the orbital motion of an electron around the atomic nucleus and (**b**) the spinning electron

moment is $\pm\,\mu_B$ (plus for spin up, minus for spin down), and the orbital spin contribution is $\mu_B m_l$, with m_l being the magnetic quantum number of the electron. Thus, the net magnetic moment of an individual atom can be approximated as the sum of the magnetic moments of its electrons. For an atom with completely filled electron shells or sub-shells (e.g. helium, neon and argon atoms), there is total cancellation of both orbital and spin moments of its electrons, and they have no macroscopic magnetic moment and are difficult to magnetise. All other materials are "magnetic" to some extent and can be classified according to the response of their individual atomic moments to an applied magnetic field, or their **magnetic susceptibility**.

1.1.2 Magnetic Susceptibility

If an external field of magnitude H is applied to a material, the individual atomic moments in the material contribute to the overall magnetic response, or **magnetic induction**, B:

$$B = \mu_0 * (H + M) \tag{1.1}$$

where μ_0 is the permeability of free space and the magnetisation, M, is the magnetic moment, μ, per unit volume V. The volumetric magnetic susceptibility, χ, describes the magnetisation induced in a material by H:

$$M = \chi * H \tag{1.2}$$

with χ being dimensionless because both M and H are expressed in Am^{-1}. The classification of materials according to their magnetic susceptibility is discussed in the next sections.

1.1.3 Diamagnetic and Paramagnetic Materials

Most materials display little magnetism, which persists only when an external magnetic field, H, is applied. These are classified either as **paramagnetic** or

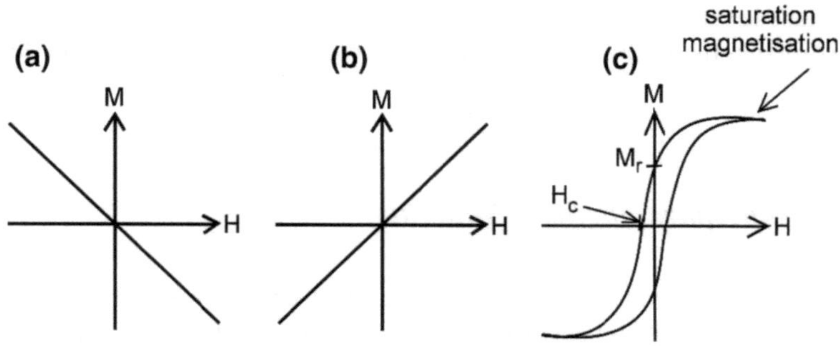

Fig. 1.2 Magnetisation, *M*, of different classes of materials in response to an external magnetic field, *H*. Note that the magnitudes of M are not drawn to scale. **a** Diamagnetic materials: the induced magnetic moment opposes the external field. **b** Paramagnetic materials: induced magnetisation aligns with and is proportional to the external field. **c** Ferromagnetic materials: characteristic sigmoidal shape of the magnetisation curve displaying a certain degree of hysteresis. M_r: remanent magnetisation at $H = 0$, H_c: coercive field required to achieve $M = 0$

diamagnetic. In diamagnetic materials, electron spins become aligned such that the direction of the overall induced magnetic moment is opposite to that of *H*, and χ is of the order of -10^{-5} (Fig. 1.2a). In paramagnetic materials the induced magnetisation aligns with *H* and χ ranges from 10^{-5} to 10^{-2} (Fig. 1.2b) [3].

1.1.4 Ferro-, Ferri- and Antiferromagnetic Materials

Some materials are magnetic even without an applied field due to long-range order of their individual, unpaired electron spin moments. These materials are classified as **ferromagnetic**, **ferrimagnetic** or **antiferromagnetic**, depending on the nature of the coupling interaction between the electrons in the material. The coupling interactions cause the spin magnetic moments of neighbouring atoms to align with each other even in the absence of an external magnetic field. This alignment can exist over large volume regions of the crystal, which are called **domains** (Fig. 1.3a). When an external field is applied, these materials can be strongly magnetised, with χ reaching values of the order of 10^5 [3]. The maximum magnetisation, or **saturation magnetisation**, is achieved when all magnetic moments in a piece of material are aligned with *H*. Unlike paramagnetic materials, the magnetisation, *M*, is not linearly proportional to *H* but displays a characteristic sigmoidal shape (Fig. 1.2c). Moreover, as *H* is reversed the curve does not retrace its original path, but displays a degree of **hysteresis**. At zero field, a residual magnetisation exists which is called **remanence**; the material remains magnetised even in the absence of *H*. To reduce the magnetisation to zero, an *H* field of a magnitude $-H_c$ must be applied in the opposite direction to that of the original field. This characteristic is called **coercivity** and *Hc* the coercive field. When ferri- or ferromagnetic materials are shrunk to the order of a few tens of nanometres, **superparamagnetic** behaviour is observed, which will be discussed in the next section.

1.1.5 Superparamagnetic Materials

Superparamagnetism is a type of magnetism that occurs in small ferro- or ferrimagnetic particles of dimensions between 3 to 15 nm [4]. These nanoparticles consist of a single domain, in which electronic spin vectors add up such that they form a total magnetic dipole moment [5]. Due to crystal structure or particle shape, there may be a particular direction along which the nanoparticle's magnetic moment aligns. If there is one preferred direction, those particles are said to have uniaxial anisotropy, and there will be two minimum energy states with anti-parallel moment orientations (Fig. 1.3b). To change from one low-energy orientation to the anti-parallel one, a certain **energy barrier**, ΔE, has to be overcome, whereby:

$$\Delta E = K * V \tag{1.3}$$

K is the anisotropy energy density and V is the volume of the particle. The direct proportionality between ΔE and V is the underlying reason for superparamagnetic behaviour in particles with small V. For these particles ΔE becomes comparable to the thermal energy at, for instance, room temperature. Therefore, thermal energy is sufficient to "flip" the magnetisation between its two preferred orientations.

Apart from the temperature, it is important to consider another factor when observing the superparamagnetic state, namely the **relaxation time**, τ, of the nanoparticle's net magnetisation. This is the average time required to perform a "flip" between the two preferred directions:

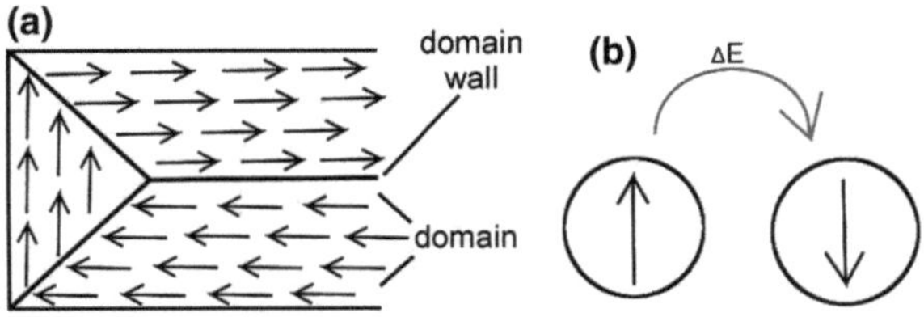

Fig. 1.3 Magnetic domains. **a** In a bulk of ferromagnetic material, spin magnetic moments (*arrows*) of neighbouring atoms align with each other in microscopic domains. Adjacent domains are separated by domain walls (*lines*), across which the direction of magnetisation gradually changes. **b** Nano-sized ferromagnetic particles (*circles*) comprise a single magnetic domain. In a nanoparticle with uniaxial anisotropy there are two minimum energy states with anti-parallel moment orientations (*arrows*), which are separated by an energy barrier ΔE

$$\tau = \tau_0 * \exp\left(\frac{\Delta E}{k_B T}\right) \quad (1.4)$$

where τ_0 is a pre-exponential factor that describes the length of time scales characteristic for the probed material (often of a magnitude around 10^{-9}–10^{-12} s), [6] and k_BT describes the thermal energy, with k_B being the Boltzmann constant and T the absolute temperature. Therefore, it has to be considered that observations of superparamagnetism are not only dependent on the temperature, but also on the measurements time, τ_m, of the experimental technique used to observe the material's magnetic properties. In typical experiments τ_m can range from slow timescales of 10^2 s for DC magnetisation to fast timescales of 10^{-8} s for ^{57}Fe Mössbauer spectroscopy [7]. If $\tau_m >> \tau$ the flipping rate is fast compared to the measurement time, such that a time-averaged net moment of zero will be observed (superparamagnetic state). But if $\tau_m << \tau$ a quasi-static, **blocked** state is observed (Fig. 1.4a). Equation 1.4 indicates a relationship between the relaxation time τ and the temperature T. A **blocking temperature**, T_B, can be defined as the temperature between blocked and superparamagnetic state, at which $\tau = \tau_m$.

Above the T_B, the net magnetic moment of an assembly of nanoparticles averages to zero. However, in an applied field there will be an alignment of magnetic moments along the external field leading to a net magnetisation. Plotting the magnetisation as a function of the applied field's magnitude, yields a magnetisation curve, to which the **Langevin function** can be fitted (Fig. 1.4b) [8]. This curve looks similar in shape to the magnetisation curve of bulk ferromagnetic materials (Fig. 1.2c), but is characterised by the absence of hysteresis, and therefore has negligible remanence and coercivity, when recorded above T_B.

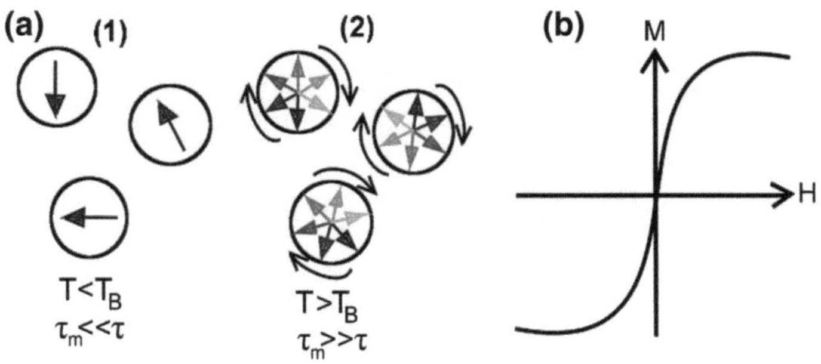

Fig. 1.4 Observing superparamagnetic behaviour. **a** The net magnetisation direction (*arrows*) of nanoparticles (*circles*) below the blocking temperature, T_B, is quasi static (*1*), whereas above T_B the moment reversals are so rapid that (if no field is applied) the time-averaged net moment is zero. **b** Magnetisation, M, of a superparamagnetic material in response to an applied field, H, recorded above T_B

1.2 Magnetic Nanoparticles

1.2.1 Nanoparticles

Nanoparticles are defined as materials with three external dimensions of equal, nanoscale magnitude, i.e. ranging from 1 to 100 nm [9]. Due to their size, nanoparticles can interact on a cellular, subcellular, or genetic scale (Fig. 1.5). In nature nanoparticles exist in abundance, e.g. as proteins, but since the late 1980s, fabrication of a variety of man-made nanoparticles has rapidly evolved [9]. These can be made from well-established materials, however, their small dimension gives rise to unique electronic, optical, and magnetic properties [10–12]. Therefore, nanoparticles have opened up new possibilities in a range of fields, e.g. in the energy sector [13] or in the biomedical arena [14]. With the rise of this new class of materials, concerns regarding their safety have been voiced [15] and continue to be monitored. In the following sections, protocols for magnetic nanoparticle synthesis and functionalisation will be briefly outlined, and aspects of potential toxicity will be discussed.

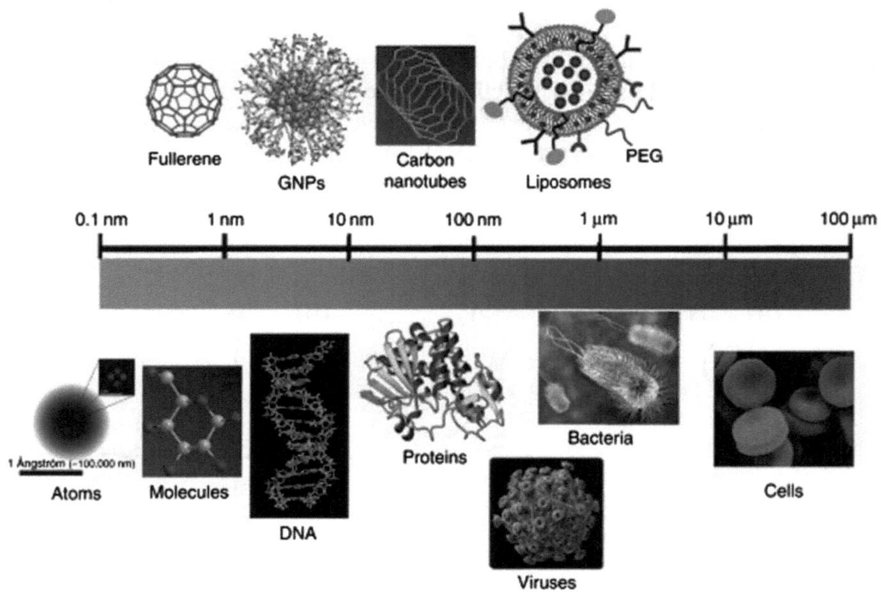

Fig. 1.5 Size of nanoparticles in relation to biological objects. Due to their small size, nanoparticles can come into close proximity with cells and cellular components, such as proteins and DNA. GNPs: glyco- nanoparticles (gold nanoparticles functionalised with carbohydrates). PEG: polyethyleneglycol (a common polymer coating to increase the circulation time of nanoparticles in the blood stream). Image re-printed with permission from Marradi et al. [16]

1.2.2 Synthesis and Functionalisation of Magnetic Nanoparticles

A wide variety of magnetic nanoparticles with different composition, size and shape can be synthesised. These include iron oxides such as Fe_3O_4 [17] and γ-Fe_2O_3 [18], pure metals, such as Co [19] and Fe [20], and alloys, such as $CoPt_3$ [21] and FePt [22] Synthesis protocols that achieve shape-controlled, stable and monodisperse magnetic nanoparticles include co-precipitation, thermal decomposition and hydrothermal synthesis [23]. Co-precipitation is a facile synthesis route, which can be carried out in water, whereas thermal decomposition takes place in organic solvents, and hydrothermal synthesis in an ethanol-water mixture. In all these methods, capping agents are added which cover the particles' surface and prevent agglomeration. This is necessary, because the small size of nanoparticles results in a large surface area and thus a high total surface energy. This is thermodynamically unstable and therefore nanoparticles tend to agglomerate to minimise the surface energy [24]. Colloidal stability of magnetic nanoparticles is either achieved through electrostatic or steric repulsion. For example, magnetite nanoparticles synthesised through co-precipitation have been stabilised electrostatically by introducing charged capping agents, [25] while nanoparticles synthesised by thermal decomposition are generally stabilised in an organic solvent by fatty acids or surfactants [26].

Superparamagnetic iron oxide nanoparticles (SPIONs) are well suited for biomedical applications because of their low toxicity [27]. Furthermore, surface coatings can be designed to add a variety of functions to the SPION (Fig. 1.6). For example, biocompatible coatings and coatings that increase blood circulation times can be used for systemic injection into animals or humans [28]. Specific ligands can

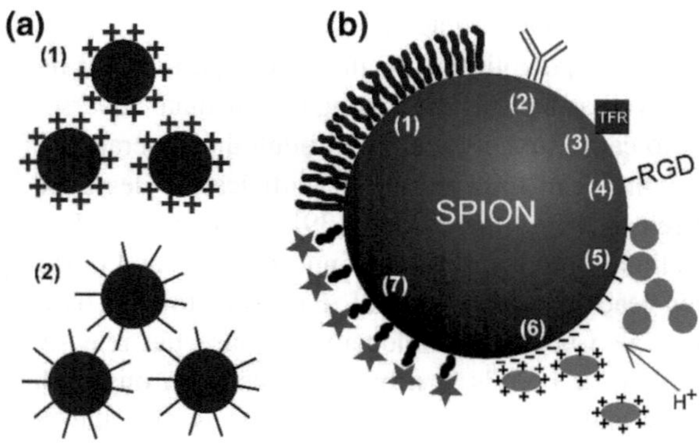

Fig. 1.6 Surface functionalisation of SPIONs. **a** Colloidal stability is achieved by electrostatic (*1*) or steric (*2*) repulsion. **b** Possible SPION coatings for biomedical applications. Polymers (*1*) confer biocompatibility. Ligands such as antibodies (*2*) transferrin (*3*) or RGD-peptide (*4*) achieve targeted delivery. Therapeutics can be covalently (*5*) or electrostatically (*6*) attached and released by specific triggers, such as pH change. Fluorescent molecules (*7*) render SPIONs a multimodal imaging probe

be attached to develop targeted magnetic nanoparticle systems, [29] and fluorescent molecules can turn them into multimodal imaging probes [30]. Finally, therapeutic agents can be covalently or electrostatically bound to the SPION surface, and stimuli-responsive release can be achieved by incorporating pH sensitive linkers [31]. Thus, magnetic nanoparticles can be customised for specific purposes which is particularly appealing for biological and biomedical applications. Some of these applications will be discussed in more detail in Sect. 1.3.

1.2.3 Potential Toxicity of Magnetic Nanoparticles

From Fig. 1.5 it is apparent that nanoparticles can interact with many components of the cell due to their small size. This opens up attractive possibilities to interact with cells on many levels, but at the same time it bears the potential to cause damage to integral parts of the cell. Although humans have been exposed to nanoparticles throughout their evolution, e.g. to carbon-based nanoparticles generated during combustion, the advent of industrialisation and the recently increasing manufacture of man-made nanomaterials have certainly added and will continue to add to the exposure burden. In 2005, Oberdörster et al. [15] proposed the establishment of "nanotoxicology" as a new discipline to systematically study effects of nanomaterials on human health and the environment. Since then, numerous studies have sought to elucidate mechanisms of nanoparticle toxicity, and aimed at developing appropriate testing strategies [32]. Nanoparticles can cause toxic effects in a variety of ways, e.g. by disrupting the cell membrane, [33, 34] interfering with cell signalling, [35] affecting protein structure and function, [36, 37] or causing DNA damage (Fig. 1.7) [38]. Therefore, toxicity assessments should always include the measurement of a variety of cell functions (often referred to as "end-points") in response to nanoparticle exposure.

Regarding the toxic potential of magnetic nanoparticles, it has been found that the metal core as well as the surface coating can mediate adverse effects. The metal core is believed to cause toxicity mainly through the generation of reactive oxygen species (ROS), which can form when the particles are degraded by the cells and metal ions are released into the cytosol [39]. An excess of ROS can lead to the destruction of cellular proteins, lipids and nucleic acids and thus impair normal cell function or even lead to cell death [40]. Exposure to iron oxide-based magnetic nanoparticles can also become problematic because the release of the metal can cause imbalances in the body's iron homeostasis and iron overload has shown to cause severe liver and heart toxicity [41]. There is also the possibility that the surface coating mediates toxicity rather than the metal core, as has been shown for some short-chain polymers [42] and poly-cationic coatings [43]. Accordingly, it is often found that different surface coatings have different effects on the cell function that is being measured [44, 45].

Several factors have been shown to affect the toxic potential of magnetic nanoparticles, the most obvious being exposure dose and time. Most studies find

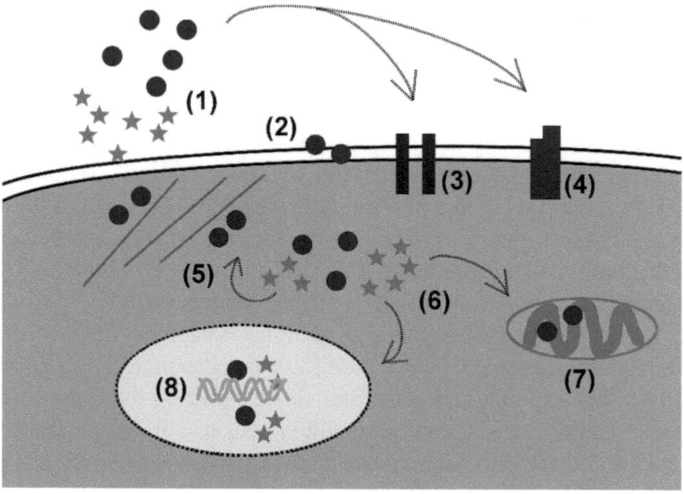

Fig. 1.7 Possible routes of nanoparticle toxicity. In the extracellular environment, nanoparticles may catalyse the formation of damaging reactive oxygen species (ROS) (*1*), disrupt the cell membrane (*2*), bind to ion channels (*3*), or cell surface receptors (*4*), which may impact on cell signalling cascades. Once the nanoparticles have been internalised, they may interfere with elements of the cytoskeleton (*5*), either directly or through the formation of ROS. Similarly, mitochondrial damage (*7*) can affect cell metabolism, and nanoparticles may penetrate into the cell's nucleus and cause DNA damage (*8*). Spheres: nanoparticles; stars: ROS

that toxic effects of magnetic nanoparticles are aggravated when cells are exposed to high doses and experience prolonged contact [46–48]. The size and shape can also influence toxic outcomes, [49, 50] which could be due to differences in uptake efficiencies and resulting differences in the effective exposure dose. Furthermore, different cell types may exhibit different responses to magnetic nanoparticles. For example, exposure to the same concentration of SPIONs has shown to cause significant toxicity in neuronal and lung cells but only little toxicity on other cell types, such as heart and kidney cells [51]. This has been attributed to differences in cell uptake, and thus again, differences in exposure burden.

1.3 Biomedical Applications of Magnetic Nanoparticles

Magnetic nanoparticles are excellent tools for a variety of biomedical applications. Being nano-sized, they can get into close contact with biological units, such as cells and proteins (see Fig. 1.5). Furthermore, their superparamagnetic behaviour (see Sect. 1.1.5) is an important prerequisite for their use in biomedicine because magnetically-induced agglomeration, which could lead to adverse effects, does not occur due to the absence of net magnetisation within the biological temperature range (around 310 K). As has been touched upon in Sect. 1.2.2, the surface of magnetic nanoparticles can be functionalised with biomolecules that can address specific structures of cells or tissues, or that have therapeutic value. For example, antibody-conjugated magnetic nanoparticles can tag specific cell types, such that they can be isolated from an heterogeneous cell populations using magnetic cell separation [52]. The capacity to immobilise or guide magnetic nanoparticles using an external magnetic field combined with the permeability of human tissue to magnetic fields also opens up the possibility for the remote control of these particles and any cargo that they may carry. Thus, cell labelling with magnetic nanoparticles has enabled remote cell manipulation [53, 54] and targeted actuation, [55–57] as well as magnetically guided drug delivery [58]. Magnetised cells can also be suspended in the culture medium using an external magnetic field, which enables scaffold-free three-dimensional cell culture [59]. Another interesting possibility lies in the ability to heat up magnetic nanoparticles by placing them in a time-varying magnetic field. Cancerous cells are particularly sensitive to temperatures in excess of 41 °C, [60] and therefore magnetically-induced hyperthermia offers a possibility to treat tumours either in conjunction with chemotherapy [61] or by thermo-ablation [62]. Finally, the great susceptibility of superparamagnetic nanoparticles to applied magnetic fields allows for applications such as contrast generation in magnetic resonance imaging, or magnetic capture and release of cells, which will be discussed in more detail in the following sections.

1.3.1 Magnetic Resonance Imaging

Magnetic resonance imaging (MRI) is a medical imaging technique, which uses non-ionising radiation. Instead, the patient is placed inside a large coil, which creates a magnetic field that is capable of aligning the magnetic moments of the body's ubiquitous hydrogen nuclei (protons) present in fluids and tissues. Broadly speaking, an MR image can be generated because different tissues have different proton compositions and therefore a different magnetic signature, which can be translated into a greyscale image. Many pathologies can thus be diagnosed, such as oedema or tumours because their fluid content is often altered compared to healthy tissue (Fig. 1.8a) [63, 64]. The images are produced using a sequence of radiofrequency and gradient pulses with carefully controlled timings and durations. The sequence timings affect image contrast, and adjustment of these timings can achieve good contrast between individual tissue types. However, some image-based diagnoses benefit from further contrast enhancement between normal and abnormal tissue. This can be achieved with contrast agents, such as paramagnetic gadolinium ion complexes or SPIONs [65] SPION-based contrast agents become magnetically saturated by the external magnetic field applied in the MRI scanner and thereby generate a significant local magnetic field, which perturbs the magnetic signal from surrounding protons. This leads to a change in signal intensity compared to neighbouring tissue that does not contain any contrast agent, so that the resulting image contrast between the two is enhanced.

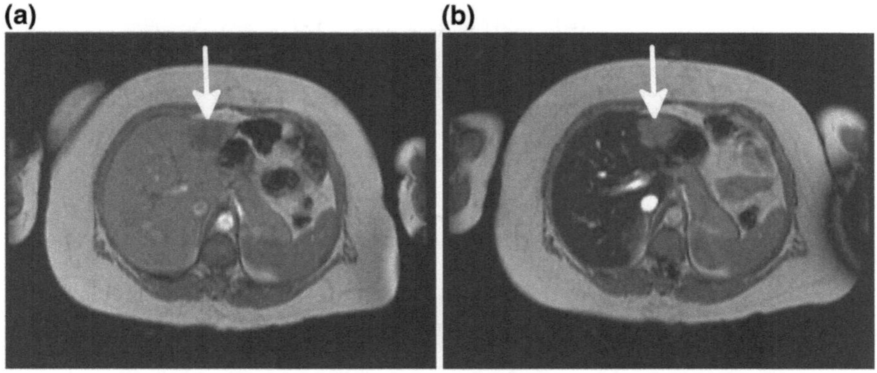

Fig. 1.8 MRI of a liver tumour. **a** Prior to SPION injection, a fuzzy dark area is visible in the liver. **b** After SPION injection, healthy liver tissue appears *dark* and the outline of the bight, abnormal tissue is clearly visible. Image adapted with permission from McRobbie et al. [66]

Depending on their size and surface functionalisation, SPIONs are selectively taken up by different tissues, such that they enhance contrast in these regions. For example, 30 nm SPIONs accumulate more rapidly in liver and spleen compared to 10 nm ones, which have a longer half-life in the blood stream and can be taken up by reticuloendothelial cells (e.g. macrophages) of lymph nodes and bone marrow [67]. Notably, the function of the reticuloendothelial system (RES) of tumour tissue is reduced compared to healthy tissue, such that circulating contrast agents are more efficiently taken up by the RES of healthy tissues compared to abnormal ones. Thus, malignant lymph nodes and tumours can be identified (Fig. 1.8b) [68, 69]. MRI can also be used to track transplanted, magnetically labelled cells in vivo with high spatial resolution, [70, 71] which helps to gain understanding of cell migration and therapeutic outcome in stem cell-based therapies (see also Sect. 1.4.4).

1.3.2 Magnetic Cell Separation

Magnetic cell separation (MACS) is a technique for the separation, concentration or removal of magnetically labelled cells from an aqueous sample by means of an applied magnetic field. With MACS, it is possible to isolate particular cell types from a mixed cell population, [52, 72] or concentrate pathogens for disease detection [73, 74].

Early magnetic separation techniques used micron-sized beads with diameters well in excess of 0.5 μm to achieve significant cell magnetisation, [75] but problems with cytotoxicity and agglomeration of cells due to magnetisation effects led to the exploration of SPIONs for these types of application [52]. The process involves conjugation of antibodies onto the SPION surface, which bind to specific cell surface markers, such that the particular cell type expressing this marker can be

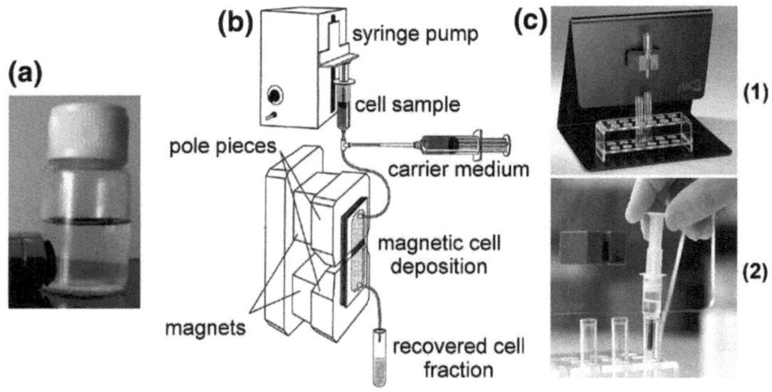

Fig. 1.9 Magnetic cell separator setups. **a** A magnet placed against a vial containing a suspension of magnetically labelled bacteria. Magnetised bacteria have agglomerated on the side of the vial, next to the magnet Image adapted with permission from from Huang et al. [81]. Copyright 2010, American Chemical Society. **b** Magnetophoretic fluid phase fractionation setup, in which magnetically labelled cells are separated from unlabelled cells because of their different magnetophoretic mobilities. Image adapted with permission from Zborowski et al. [82]. Copyright 1995, American Chemical Society. **c** Magnetic cell separation using columns packed with steel beads. The columns are placed in a permanent magnet (*1*), such that magnetically labelled cells are immobilised. To elute the trapped cells, the column is removed from the magnet and a plunger is used to push the cells out with a volume of elution buffer (*2*). Images (*1*) and (*2*) taken from the Miltenyi Biotec website (www.miltenyibiotec.com)

separated from a mixture of cells using a magnet. Therefore, it has been possible to specifically magnetise cancer cells, [76, 77] particular stem cell populations, [78] or bacteria [79] using antibody conjugated magnetic particles. Although antibody conjugation imparts specificity to MACS, this specificity may not always be required. Simpler surface functionalisations of the SPION surface, such as small molecules or charged functional groups, can be very effective in binding SPIONs to bacterial surfaces or chelating toxic metal ions, such that these contaminants can be removed using a magnetic field [74, 80].

Magnetic separators can be as simple as a permanent magnet placed against a test tube to cause agglomeration of the magnetically tagged cells in suspension, followed by the removal of the supernatant (Fig. 1.9a) [81]. However, this method may suffer from slow agglomeration rates, which reduce separation efficiency. Magnetic separator designs that produce high magnetic field gradients can increase separation efficiency by trapping magnetised cells as they flow through a channel or a column. For example, magnetically labelled cells can be separated from non-labelled cells using a fluid phase fractionation setup. A cell suspension is passed through a region to which a magnetic field gradient is applied, and the cells are separated because of their different magnetophoretic mobilities (Fig. 1.9b). In a simpler setup, a column is packed with a magnetisable matrix, e.g. steel beads, and the cell suspension is passed through it while a magnetic field is applied. The close-packing of the steel beads means that magnetically labelled cells passing through the column are always close to high magnetic field gradients that immobilise and concentrate them in the column. When the magnet is removed, the cells demagnetise due to the superparamagnetic nature of the SPIONs, such that they can be eluted from the column (Fig. 1.9c). Using this method, a sample can either be depleted from a magnetically labelled cell population (negative selection), or the magnetically labelled cells can be isolated from a mixed population (positive selection) and used for subsequent culture of this specific cell type [52].

1.4 Stem Cells

Many excellent textbooks and review articles on stem cell biology and applications of stem cells in regenerative medicine are available. Books by Lanza [83] and Stocum [84] give a thorough overview over these topics, and reviews by Fuchs et al. [85] Wu et al. [86] and Tabar et al. [87] cover specialised topics such as the stem cell niche and recent progress in regenerative medicine.

1.4.1 Characteristics and Classification

Stem cells are characterised by their capacity to perform numerous cell division cycles that result in the formation of identical daughter cells (self-renewal), and by their ability to give rise to a variety of specialised cell types (differentiation) [88]. The "original" stem cell of each individual is the zygote, or fertilised egg. From the zygote, the embryo first develops into a mass of **pluripotent embryonic stem cells**, which have the capacity to differentiate into any specialised cell type of the adult individual. As development progresses, these embryonic stem cells develop into more restricted **foetal stem cells** that give rise to particular cell types that eventually make up tissues and organs. A reservoir of these stem cells is retained in the tissues after birth to enable repair and regeneration following injury or disease. These **adult stem cells** share some characteristics with their embryonic predecessors, in that they have the capacity for self-renewal and differentiation. However, they are not pluripotent anymore, they are **multipotent**, i.e. they can differentiate into particular cell lineages. Examples of adult stem cells are haematopoietic stem cells, which give rise to red and white blood cells, [89] and mesenchymal stem cells, which were used in this study and are discussed in more detail in the next section.

1.4.2 Mesenchymal Stem Cells

Mesenchymal stem cells (MSCs) are multipotent stem cells that can be found in several adult tissues, such as bone marrow and adipose tissue [90, 91]. Human MSCs (hMSCs) can be identified and isolated based on the presence (or absence) of

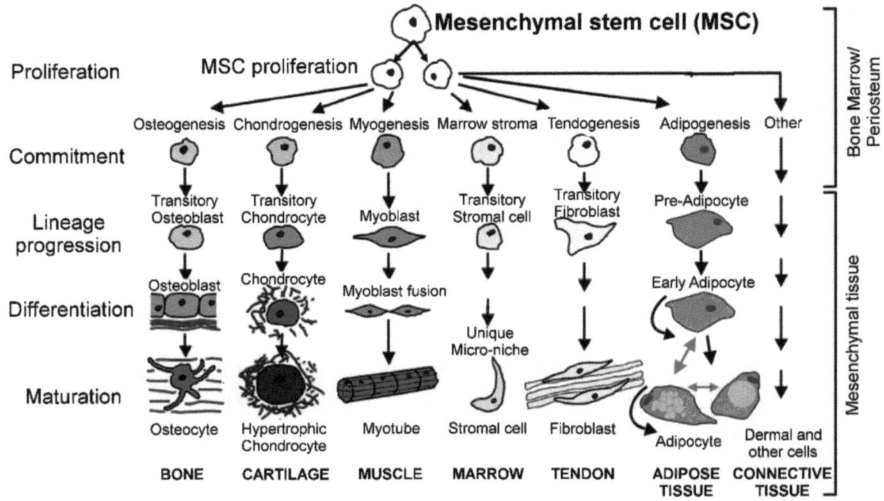

Fig. 1.10 Differentiation pathways of mesenchymal stem cells (MSC). MSC have the capacity to proliferate and differentiate into different cell types depending on the environmental conditions. Image adapted with permission from DiMarino et al. [100]

specific surface markers [92]. Isolated hMSCs can be kept in an in vitro culture where they retain their differentiation capacity [93] and only reach senescence after approximately 15 population doublings [94]. Cultured hMSCs can differentiate into a variety of cell types, such as bone, [95] cartilage, [96] fat [97] and muscle cells [98], given specific biochemical or mechanical cues [99, 100] (Fig. 1.10). The differentiation pathway includes a series of pre-programmed, genetically regulated events, including lineage commitment, lineage progression and eventual differentiation and maturation of the specialised cell type [88].

1.4.3 *hMSC-Based Therapies*

Regenerative medicine aims at harnessing the capacity of stem cells to regenerate and rejuvenate tissues [101]. The foundation for this branch of medicine reaches back to the early 1980s and the discovery that embryonic stem cells can be derived from the inner cell mass of mouse embryos, both in vivo and in vitro [102, 103]. Human embryonic stem cells were first cultured in vitro in 1998, [104] however, the use of embryonic stem cells has faced ethical controversy, [105] and transplantation of these cells bears the risk of immune system rejection and teratoma formation (i.e. formation of a tumour composed of tissues not normally present at the site) [106]. Adult stem cells, such as hMSCs, are a good alternative to embryonic stem cell therapy, because they are less ethically controversial and it is possible to derive autologous cells (i.e. from the patient's own body), which reduces the risk of immune rejection.

hMSCs have the potential to repair a variety of damaged tissues, either by direct injection or by engineering functional tissue constructs that can be implanted into an injured area. Locally or systemically injected hMSC can regenerate damaged tissue either directly by differentiating into rejuvenated tissue, or indirectly by secreting cytokines that reduce inflammation and apoptosis (i.e. programmed cell death) [107]. Thus, hMSCs have been shown to regenerate damage due to stroke [108] and myocardial infarction, [109] and alleviate symptoms of Parkinson's disease [110]. It is also possible to grow MSCs on scaffolds and differentiate them in vitro into particular tissue types that can be implanted. This tissue engineering approach has been used to heal bone defects [111], osteoarthritic cartilage damage [112] or even replace a main bronchus [113].

1.4.4 Magnetic Nanoparticles in MSC-Based Therapies

Although MSCs are promising candidates for regenerative medicine interventions, their migration, survival, integration into newly formed tissue and eventual fate are poorly understood. Therefore, a non-invasive method for tracking implanted MSCs in animal models or human trials is essential to understand and predict therapeutic outcomes. MRI is a good method to achieve this, because it offers high spatial resolution, uses non-ionising radiation and can detect the migration of transplanted cells over time [114, 115]. MRI tracking requires that implanted MSCs are labelled with a contrast agent, and SPIONs are good candidates due to their low toxicity and their large contrast enhancing effect. Remarkably, single cells labelled with SPIONs can be tracked in vivo using MRI, [71] and tracking of SPION-labelled MSCs is possible over several weeks [116] or even months (Fig. 1.11) [117].

Incorporating magnetic nanoparticles into stem cells also enables their remote manipulation. For example, magnetic fields can be used to guide magnetically tagged stem cells to particular sites [118]. Furthermore, magnetic nanoparticles can activate stem cell differentiation through mechanical stimulation. For this application, SPIONs are functionalised with antibodies or peptides that can attach to mechanosensitive cell surface receptors or ion channels. By applying a magnetic field, a force can be exerted on the particles and consequently on the receptors, which results in receptor activation and a subsequent cell signalling cascade in MSCs [119]. Thus, it has been shown that bone formation can be promoted in MSCs tagged with magnetic nanoparticles and subjected to an external magnetic field, both in vitro and in vivo [56, 120].

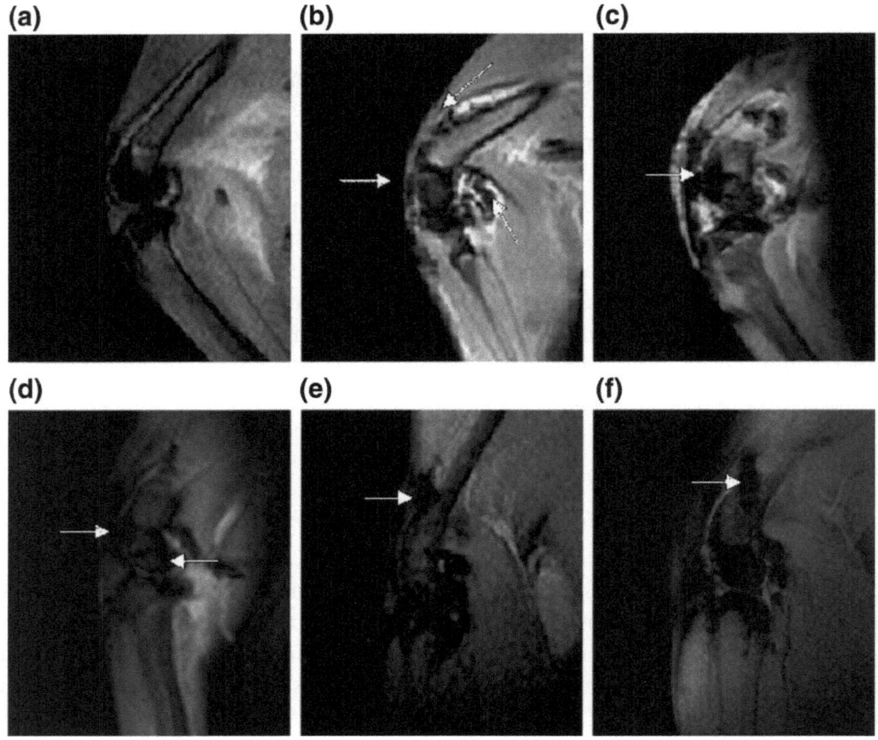

Fig. 1.11 Tracking of SPION-labelled MSCs transplanted into a rabbit's knee joint with an induced cartilage defect. **a** Image obtained before MSC injection. **b** Image obtained within 1 h of injection into the knee joint. Areas of low signal intensity (i.e. *dark spots*) indicate the presence of SPION-labelled MSCs. **c**, **d** Images obtained 4 and 8 weeks after injection. **e**, **f** Images obtained 12 weeks after injection. Reproduced from Jing et al. [117]. Copyright 2008. Elsevier Masson SAS. All rights reserved

1.5 Bacteria

Bacteria are a large group of single-celled organisms, which lack a membrane-enclosed nucleus and mitochondria, and are typically 1–8 μm in length. Many bacterial species live commensally in humans (i.e. they profit from living in/on humans without affecting them), [121] or provide health benefits, e.g. by aiding digestion [122] or preventing growth of harmful bacteria [123]. However, there are also pathogenic species that can cause a variety of life-threatening infectious diseases, such as cholera or tuberculosis [124]. Bacteria can be divided into two major groups: Gram positive and Gram negative. The distinction between both groups is based on the Gram stain (crystal violet), which stains bacteria based on the composition of their cell wall [125]. An excellent introduction into the biology of bacteria can be found in Brock Biology of Microorganisms [126].

1.5.1 *Gram-Positive Bacteria*

Gram-positive bacteria can be identified by Gram staining, because their cell wall retains the crystal violet stain in its thick peptidoglycan layer. This peptidoglycan layer covers the cytoplasmic membrane and contains a variety of acidic, carbohydrate-based polymers, such as teichoic acids, that play important roles in cell adhesion [127] and immune activation (Fig. 1.12a) [128]. Gram-positive bacteria lack the outer membrane/lipoplysaccharide (LPS) layer that is characteristic of gram-negative cells and that is associated with mediating pathogenicity. Therefore, fewer gram-positive species act as pathogens. The absence of the outer membrane also makes them more susceptible to antibiotics, however, they possess drug efflux pumps which mediate antibiotic resistance [129].

1.5.2 *Gram-Negative Bacteria*

Gram-negative bacteria do not retain the crystal violet stain that is used in Gram staining. The gram-negative cell wall is a multi-layered structure, in which a thin peptidoglycan layer lies between an inner and outer membrane (Fig. 1.12b) [130]. During Gram staining, an alcohol wash is applied which de-colourises the cells because their peptidoglycan layer is too thin to retain the stain. A counterstain is added that re-colourises the bacteria, such that they can be visualised [131]. The outer membrane of gram-negative bacteria contains lipopolysaccharides, which are the main mediators of pathogenesis because they elicit a strong immune response that leads to inflammation [128]. Gram-negative bacteria are in general more resistant to antibiotics compared to gram-positive bacteria [132]. This has been

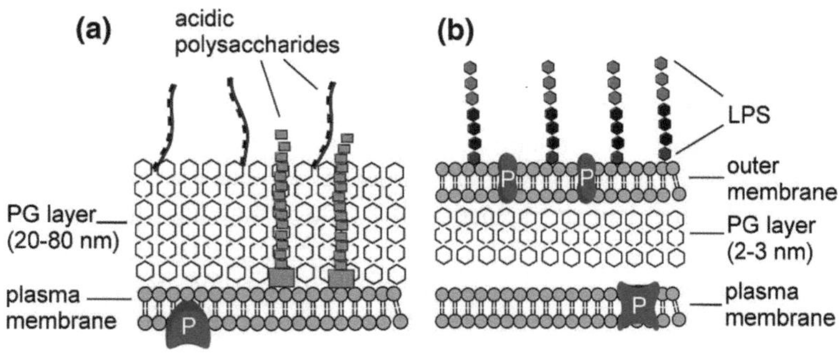

Fig. 1.12 Bacterial cell walls. **a** Gram-positive cell walls are characterised by their thick peptidoglycan (PG) layer covering the plasma membrane and containing acidic polysaccharide chains, such as teichoic acids. **b** Gram-negative cells walls are characterised by their outer membrane containing lipopolysachharides (LPS), and a much thinner PG layer. P: membrane-bound proteins

partly attributed to their outer membrane, which acts as a permeability barrier, [133] as well as to the presence of multiple drug efflux pumps, which move antibiotic molecules out of the cell [134].

1.6 Cell Labelling with SPIONs

1.6.1 Internalisation and Long-Term Fate

SPIONs are widely used in biological applications due to their favourable magnetic properties and their low toxicity [27]. SPIONs can be used in vivo, in which case they are either injected systemically, e.g. for MRI contrast, or locally, e.g. for hyperthermia treatment of tumours. They can also be used to magnetically label cells in vitro for applications such as MACS or cell tracking with MRI. For in vivo applications, it is important to understand the fate of injected SPIONs, and in vitro magnetic labelling relies on achieving adequate cellular loading with SPIONs. Therefore, many studies have been concerned with the uptake mechanism in vitro and long-term fate of SPIONs in vivo.

Cellular uptake of nanoparticles in general is believed to occur predominantly via endocytosis. Endocytosis is a form of active transport, by which cells internalise molecules and particles from their surroundings. Under its broad definition, endocytosis includes a variety of uptake mechanisms, which include phagocytosis ("cell eating"), pinocytosis ("cell drinking"), clathrin-dependent (or receptor-mediated) endocytosis and clathrin-independent endocytosis (Fig. 1.13). Phagocytosis is mainly conducted by specialised cells, e.g. macrophages, which engulf large solid particles (> 700 nm) such as pathogens [135]. Most other cell types internalise smaller particles ranging from a few to several hundred nanometres through pinocytosis [136].

Clathrin-mediated endocytosis has been identified as the primary mechanism for the cellular uptake of nanoparticles (including SPIONs) but recently a range of clathrin-independent pathways have also been identified [137]. Significantly, it has been shown that the physico-chemical properties of nanoparticles (e.g. size, shape, surface functionalisation) influence the endocytotic pathway by which they are internalised [138, 139]. For SPIONs in particular, cellular internalisation can also be manipulated using applied magnetic fields, which may alter the internalisation pathway [140].

Clathrin-mediated endocytosis can be broken down into the following steps: contact of the nanoparticle with the cell surface, cell membrane invagination, formation of a clathrin-coated pit (clathrins being proteins that assemble into a net like basket around the membrane invagination), detachment of the newly formed, membrane-bound endocytotic vesicle (endosome), and transport of this vesicle into the cytosol [141].

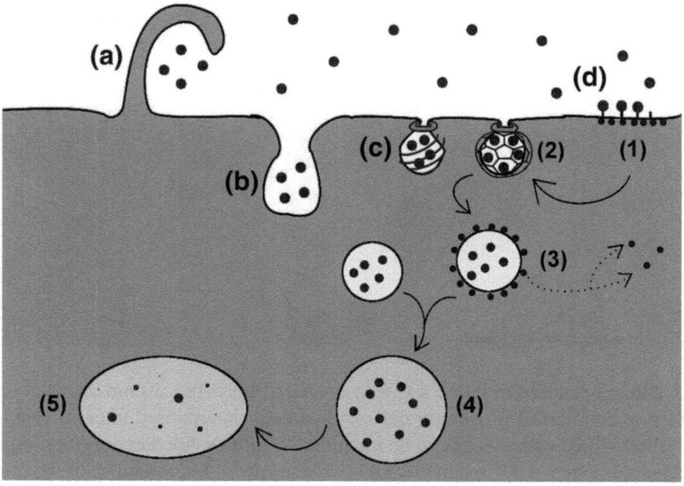

Fig. 1.13 A selection of possible endocytosis pathways for the internalisation of SPIONs. **a** Macropinocytosis characterised by membrane protrusions enveloping particles from the fluid phase. **b** Pinocytosis characterised by membrane invagination and uptake for particles from the fluid phase. **c** Caveolin- dependent endocytosis; the vesicle is coated with the protein caveolin. **d** Clathrin-dependent endocytosis. Particles bind to receptors on the cell surface (*1*), and a clathrin molecules form a net-like basket around the membrane invagination (clathrin coated pit) (*2*). Dynamin pinches the vesicle off the membrane and the coated vesicle is transported into the cytosol (*3*). Here, clathrin is recylced back to the cell membrane and the vesicle fuses with other endosomes (i.e.membrane-bound compartments) to form larger (late) endosomes (*4*). These may fuse with lysosomes (*5*), in which particles may be degraded

For charged nanoparticles, internalisation via macropinocytosis seems to be an important mechanism [138]. During macropinocytosis, extracellular fluid and its contents are captured by membrane protrusions and internalised into cells in large, heterogeneous vesicles [142]. It has been shown that anionic SPIONs are mainly located on the cell surface after a 15 min exposure (Fig. 1.14a), and can be found inside the cell in large vesicles after 1 h exposure (Fig. 1.14b). Once internalised, SPION-filled vesicles can fuse with one another and subsequently with lysosomes, which contain varying amounts of hydrolases leading to the degradation of internalised particles [143]. It has been suggested that the released iron is then stored in intracellular ferritin [144]. SPION-filled vesicles have been shown to be transferred to daughter cells during mitosis (Fig. 1.14c), [145] indicating that cellular iron content is further and further diluted through cell division.

Regarding the long-term fate of SPIONs in vivo, a recent paper has shed light on the morphological and structural changes of gold nanoparticles coated with iron oxide up to one year after their injection into mice [146]. Most nanoparticles were localised in the liver and spleen of the animals, and the degradation of the iron oxide layer over time was dependent on the surface coating. Dissolution of the iron oxide coating in the spleen was observed from day 14 post-injection, and one year after administration more than 10% of polymer-coated iron oxide still persisted in

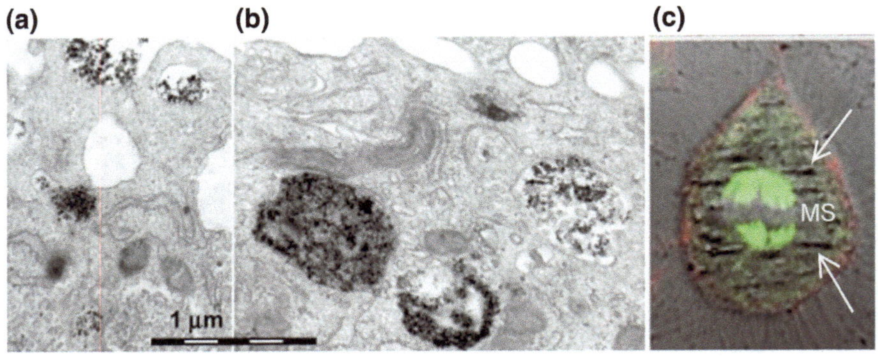

Fig. 1.14 Cellular uptake of SPIONs. **a** Citrate-coated SPIONs (ca. 30 nm diameter) can be found on the surface of endothelial progenitor cells and partially internalised after 15 min. **b** After one hour large SPION-filled endosomes can be identified. **c** SPION-filled endosomes aligned inside a HeLa cell after applying a 0.1 T magnetic field (white arrows). Microtubules of the mitotic spindle (MS) are stained with a green fluorescent dye. SPION-filled endosomes can be seen on both sides of the MS, indicating that they will be distributed equally between daughter cells after cell division is complete. Image adapted with permission from Wilhelm et al. [145]

the liver and 2% in the spleen. Significantly, it was shown that the elimination of iron was due to local, intracellular transformation and dissolution rather than excretion of intact particles from the organ.

1.6.2 Strategies to Enhance Internalisation

Cell labelling with SPIONs can be achieved by attachment to the cell surface, [52, 55] or internalisation of the particles via endocytosis. Many endocytosis pathways require that SPIONs make contact with the cell membrane before being internalised. Therefore, internalisation efficiency of SPIONs is greatly influenced by cell-particle interactions, [147] which depend on cell membrane characteristics and SPION surface properties such as charge, hydrophobicity, or the presence of targeting ligands. In addition to these interactions at the bio-nano-interface, other features such as nanoparticle size, shape and roughness also influence cell uptake [148, 149]. In general, weak cell membrane-SPION-interactions lead to low internalisation efficiencies that can be compensated for by increasing incubation times or SPION concentrations [150]. However, these conditions can adversely affect cell viability, [46, 47] and cell types with low internalisation capacity, such as lymphocytes, exhibit limited SPION uptake even after prolonged incubation times (i.e. 40 h) [151]. Alternatively, external magnetic fields can be used to enhance SPION uptake. External fields can accelerate the sedimentations of SPIONs on the cell surface and thereby facilitate the first step required for cell uptake, [152] or the

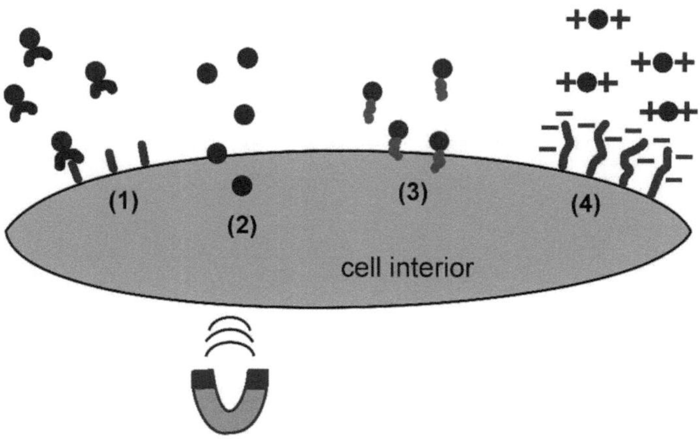

Fig. 1.15 Strategies to enhance cellular uptake of SPIONs. Targeting molecules can promote receptor-mediated endocytosis (*1*), external magnetic fields can enhance internalisation either by aiding sedimentation of SPIONs on the cell surface or by pulling the particles directly through the cell membrane (*2*), cell penetrating peptides can shuttle SPIONs through the cell membrane (*3*), and charge functionalisation leads to electrostatic adsorption of SPIONs to the cell membrane (*4*)

magnetic force can pull SPIONs directly through the cell membrane (Fig. 1.15) [140, 153].

Chemical functionalisation of the SPION surface is a strategy that has been extensively used to increase cellular uptake and to reduce labelling times to a few hours. A selection of SPION functionalisations and specifically their cellular uptake efficiency in stem cells has been summarised in Table 1.1. For instance, targeting molecules, such as antibodies, are used to promote receptor-mediated endocytosis [46, 154]. Transfection agents, such as lipids and dendrimers, [43, 155–157] or conjugated cell-penetrating peptides [158–161] have also been shown to significantly enhance labelling efficiency. However, antibody and peptide conjugations are limited by complex synthetic preparation and safety concerns over the use of xenogeneic molecules (e.g. in the case of HIV-derived tat peptide), while transfection agents can induce nanoparticle precipitation [162] and may adversely affect cell function [43]. Accordingly, there is a need for a simple surface functionalisation methodology that can produce a biocompatible SPION capable of rapidly magnetising a variety of cell types.

The ideal SPION would achieve high intracellular iron loading within a short incubation time and using low concentrations of extracellular SPIONs. These conditions are also likely to circumvent toxic effects due to prolonged SPION exposure or elevated SPION concentration. Charge-functionalisation of SPIONs with for example citrate ions or a silica shell has yielded anionic nanoparticles that are internalised very efficiently via adsorption mediated endocytosis (Fig. 1.15) [167, 168]. Thus, short incubation times of charge-functionalised nanoparticles in the range of 20 min to one hour have shown to achieve sufficient magnetisation of a variety of cell types for use in MRI [145, 164]. Cationic charge-functionalisation

Table 1.1 A selection of previously reported SPION functionalisations, incubation times and cellular iron contents

SPION functionalisation	Exposure time (h)	Dose (μg mL^{-1})	Cell type	Cellular Fe (pg $cell^{-1}$)	References
Dextran	24	50	Human MSC	0.2	Arbab [163]
	2	250	Human HPC	2.5	Daldrup-Link [46]
	0.5	108	Human MSC	0.5	Suh [160]
Dendrimers (MD-100)	48	25	Rat OPC	9	Bulte [155]
Dendrimers (Superfect)	2	25	human MSC	30	Frank [156]
Cell penetrating peptide (tat)	1	2–40	Mouse NPC	8-30	Lewin [159]
Poly-(L-lysine)	24	25	Human MSC	20	Mailander [150]
	2	25	Human MSC	16	Frank [156]
Protamine sulphate	24	50	Human MSC	10	Arbab [163]
	0.5	108	Human MSC	2	Suh [160]
Carboxydextran	24	25	Human MSC	3	Mailander [150]
Citrate	0.5 (+ 1 h chase)	28 - 560	Human EPC	2-10	Wilhelm [164]
Transferrin	2	250	HUMAN HPC	10	Daldrup-Link [46]
Liposome	24	20	Mouse HPC	14	Schwarz [157]
	2	100	Human HPC	2	Daldrup-Link [46]
	2	25	Human MSC	8	Frank [156]
Cationic polymer (PDMAAm)	72	15	Human MSC	37	Babic [165]
Cationic polymer (PEI-SA)	1	3.5	Rat MSC	50	Guo [166]

Legend: *MSC* mesenchymal stem cell, *HPC* haematopoietic progenitor cell, *OPC* oligodendrocyte progenitor cell, *NPC* neural progenitor cell, *EPC* endothelial progenitor cell, *PDMAAm* poly-(N, N'-dimethyl acrylamide), *PEI-SA* polyethylenimine-stearic acid

using polymers can achieve even higher iron loading into MSCs within 1 h incubation, [166] presumably due to the fact that the cell surface contains a multitude of anionic domains that enable adsorption-mediated endocytosis of cationised SPIONs. However, cationic SPION-functionalisation mainly employs the conjugation of polycationic transfection agents, most of which induce cytotoxic effects even at low doses, [43, 169] and are therefore not approved for clinical use (with the exception of protamine sulphate) [170].

1.6.3 Bacterial Cell Labelling

Magnetic labelling of bacteria enables the extraction of pathogens from aqueous samples, which offers the possibility to recover reporter microbes from a test environment, [171] or to isolate and pre-concentrate pathogens from large volumes [74]. This is an important step in microfluidics-based diagnostic procedures, because clinical samples of blood or urine have a volume of around 10–20 mL but microfluidic devices can only handle 50 to 200 μl. To this end, SPIONs have been functionalised with Vancomycin, a small molecule probe that targets the bacterial cell wall of gram-positive cells, and to some extent the gram-negative cell surface [74, 172]. Thus, it has been shown that several bacterial species could be magnetically captured even at very low concentrations and detected either using scanning electron microscopy [172, 173] or fluorescence microscopy, [174] and identified using mass spectrometry [175]. However, this application does not necessarily require the specificity that Vancomycin offers, but can also be achieved by charge-functionalising SPIONs with amine- and carboxyl- groups [80, 81].

1.7 Ferritin

Ferritin is a fascinating protein and many excellent reviews give a thorough insight into its biology and biochemistry, e.g. Harrison et al. [176] and Theil et al. [177]. A brilliant summary of technological applications of ferritin is given by Uchida et al. [178].

1.7.1 Structure and Function

Iron is poorly soluble in aqueous environments at physiological pH and highly toxic due to its ability to form reactive oxygen species [179]. Nevertheless, it is a vital element for almost every living organism. Therefore, a means for storing as well as solubilising iron is essential, and iron storage proteins like ferritin and ferritin-like proteins are ubiquitous among bacteria, plant and animal species [176]. Although there are considerable differences in the amino acid sequence, the overall structure and function of ferritin is highly conserved across tissues and species [180].

The outer diameter of mammalian ferritin is approximately 12 nm, and the entire protein consists of 24 subunits arranged in octahedral 432 symmetry. The cavity they enclose has a diameter of 6–8 nm and is connected to the exterior via 3–4 Å wide channels formed along the threefold cage symmetry axes [181]. An iron-rich core of up to 7 nm in diameter can be mineralised inside the cavity. If no core is present, the protein is referred to as apoferritin (ApoF) (Fig. 1.16).

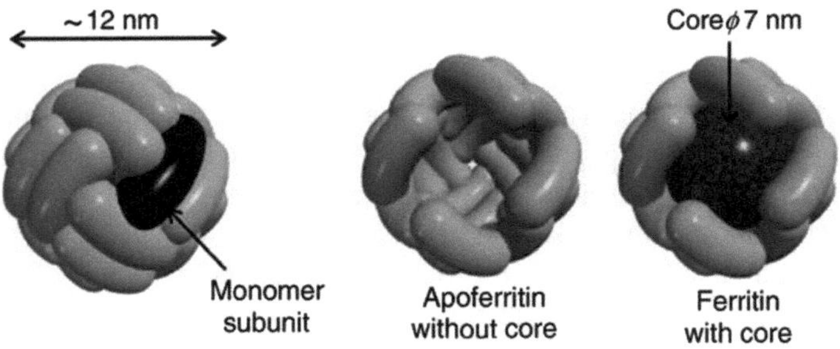

Fig. 1.16 Schematic of mammalian ferritin. The protein is composed of 24 monomer subunits and has an outer diameter of approximately 12 nm. The subunits enclose a cavity, in which an iron-rich core can be mineralised. The "empty" ferritin cage is referred to as apoferritin. Image re-printed with permission from Yamashita et al. [182]

Ferritin subunits may be classified as heavy (H) or light (L), depending on their amino acid sequence. The ratio of H to L units of ferritins varies between tissues and species, and affects the capacity to incorporate iron in the ferritin cavity. Generally, L-rich ferritins are found in iron storing organs such as liver and spleen. These ferritins contain a relatively high number of iron atoms (1500 atoms/molecule or more). H-rich ferritins are found in heart and brain, and have a relatively low average iron content (less than 1000 iron atoms/molecule) [183]. Up to 4500 iron atoms can be accommodated in the ferritin cavity, and different species can adjust the ratio of H- to L-chains in their tissue specific ferritins to meet individual needs [184]. Each subunit contains a 5 nm long four- α-helix bundle with a fifth α-helix across one end and a loop on the other (Fig. 1.17a). Twelve pairs of subunits bind "head to foot" to form the entire 24-mer protein (Fig. 1.17b). This hierarchical composition results in great stability towards heat and acids [185].

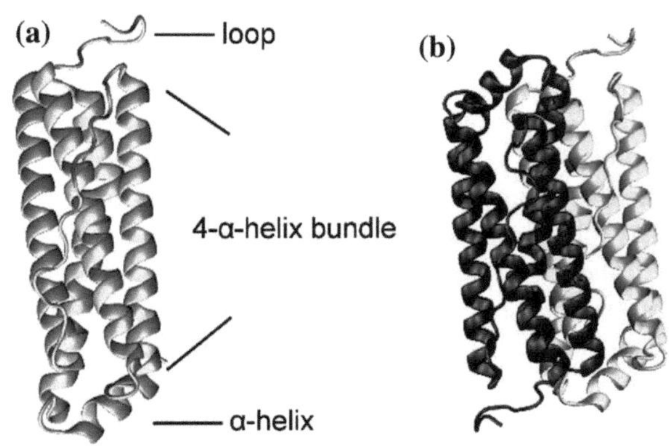

Fig. 1.17 Ferritin subunit composition. A representation of the human heavy chain subunit (PDB reference: 2FHA) was chosen to illustrate the protein structure. **a** The subunit is composed of a four- α-helix bundle with a fifth α-helix across one end and a loop on the other. **b** Subunits bind "head to foot" and twelve of these pairs make up the 24-mer protein

1.7.2 Mineralisation of the Native Ferrihydrite Core

The process of bio-mineralisation of ferric oxides inside the ferritin cavity varies slightly between species, [177] but is thought to be mainly electrostatically driven. In mammalian cells, Fe(II) ions accesses the ferritin cavity via hydrophilic channels formed between subunit interfaces (Fig. 1.18a) [186, 187]. The channels are lined with negatively charged amino acid residues which cause an electrostatic potential that guides the positively charged metal ions towards the inside of the protein cage (Fig. 1.18c) [188]. Ferroxidase sites are located at the exit of these channels on the inside of the cavity. Here, Fe(II) is oxidised to Fe (III) in the presence of molecular oxygen through a reaction that can be represented as:

$$4\,Fe^{2+} + O_2 + 6\,H_2O \rightarrow 4\,Fe\,(O)\,OH + 8\,H^+$$

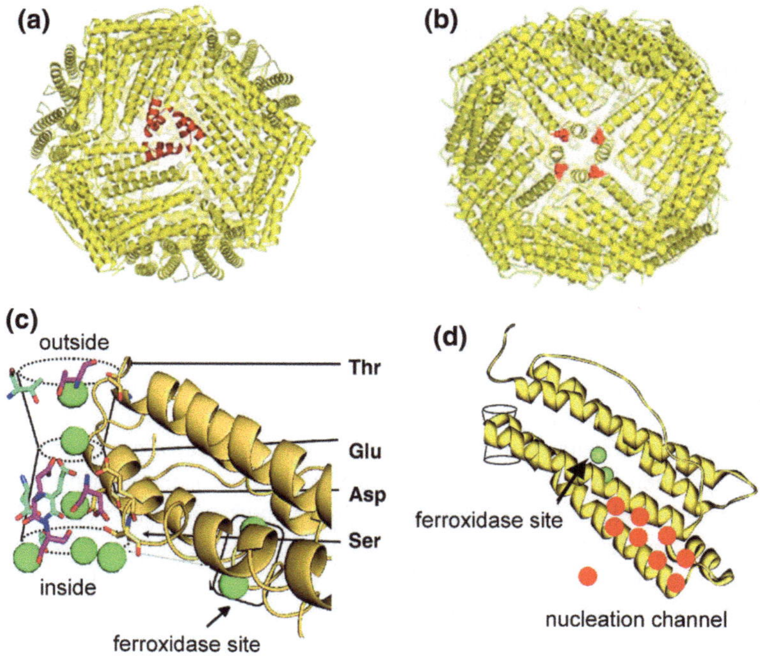

Fig. 1.18 Mechanism of iron mineralisation in the ferritin cavity. **a** Outside view of the entry to an hydrophilic channel around the three-fold axis. The sections of the protein subunit structures that form part of the channel are highlighted in *red*. **b** Inside view of a ferritin protein cage, the four-fold axis is indicated in *red*. The protein cage exerts control over mineral growth through clustering of nucleation sites around this four-fold symmetry axes. **c** Close up of an iron entry channel and nearby ferroxidase site. The blue and pink stick figures represent aspartic acid residues from the two neighbouring subunits. (D) Close-up of a nucleation channel that receives Fe(III) precursors from the ferroxidase sites and transports them to nucleation sites. Green spheres: Fe(II) ions; *red spheres*: Fe(III) mineral precursors. Thr: threonine, Glu: glutamic acid, Asp: aspartic acid, Ser: serine. Image adapted with permission from Theil [181]

It is believed that Fe(III) precursors move through nucleation channels between the ferroxidase sites and nucleation sites (Fig. 1.18d) [189]. Mineralisation of the ferrihydrite core is facilitated by those nucleation sites, which are located at the four fold symmetry axes (Fig. 1.18b). Thus, nucleation occurs oriented by four sites, leading to the formation of a highly ordered ferrihydrite core [Fe(O)OH] whose growth is limited to 5–7 nm by the size of the surrounding protein cage [190]. Ferroxidation sites are mainly situated on H- subunits, whereas L-subunits are catalytically inactive [176]. The L-subunits have a greater negative charge on the interior surface, which forms part of the nucleation sites (mainly containing aspartic acid and glutamic acid) [191]. In the ferrihydrite core, iron is completely sequestered and rendered inert [192].

1.7.3 Ferritin in the Cellular Iron Metabolism

In humans, approximately 1 mg of iron is absorbed daily from the diet, yet 20–22 mg of iron cycle into and out of the red blood cell pool [193]. Thus, most of the body's iron is located in the blood, such that blood loss results in a rapid depletion of iron. Replacement does not come from intestinal absorption, which would be insufficient given the low amounts that are taken up daily, but rather from ferritin stores.

Initially, iron is absorbed from the diet through the intestine and carried throughout the body chelated to the serum protein transferrin [194]. Iron-carrying transferrin binds to specific transferrin receptors on the cell surface and is internalised via receptor-mediated endocytosis [195]. Iron is released as soluble Fe(II) from transferrin through acidification in the endosome and eventually transported into the cytosol. There, it is available to iron-utilising proteins, and sequestered by ferritin for storage (Fig. 1.19). Cellular iron homeostasis is then regulated by the coordinated production of transferrin receptors and ferritin: cells requiring more iron up-regulate the expression of transferrin receptor and down-regulate ferritin expression [196]. Conversely, when excess iron is present, the expression of transferrin receptors is decreased and levels of ferritin increase. Cellular uptake of SPIONs has shown to elicit this response, too, with transferrin expression being decreased and ferritin expression increased after exposure [197]. How iron is mobilised from ferritin is largely unclear, but it has been proposed that this occurs mainly through transfer of ferritin to lysosomes and subsequent lysosomal proteolysis [198].

1.8 Magnetoferritin

As discussed in Sect. 1.7.2, ferritin takes up Fe(II) and forms nano-sized ferric oxide mineral cores inside the protein cavity. In vitro, ApoF can incorporate a variety of metal ions such as Pd, [199, 200] Ag, [201] Ni, [202, 203] Co, [202] Cu, [201] Cr, [203] and Mn, [204] and direct the formation of several metal oxide

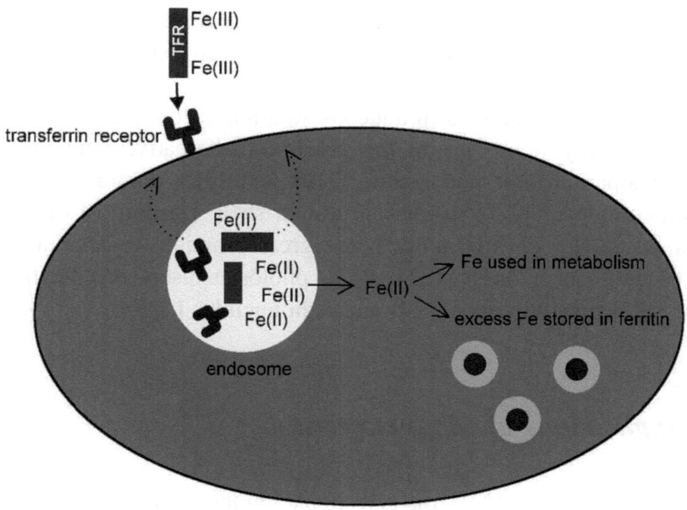

Fig. 1.19 Schematic of cellular iron metabolism. Iron is absorbed from the diet and transported as Fe(III) throughout the body chelated to transferrin (TFR). Cells can take up iron-loaded transferrin *via* receptor mediated endocytosis. In the acidified environment of the endosome, iron ions, transferrin and the transferrin receptor dissociate. Transferrin and the transferrin receptor can be recycled, and Fe(II) ions are released into the cytosol, where they can be utilised in cell metabolism or stored in ferritin

nanoparticles [178, 205, 206]. Uchida et al. [178] suggested that the surface charge directed mineralisation inside ApoF cages may explain ferritin's low specificity for iron in vitro, and consequently the possibility to mineralise other metal cations. The first synthetic mineralisation of a maghemite core inside the ApoF cavity has been named magnetoferritin (MF) [207, 208]. Since then, a variety of magnetic cores based on metals such as Ni, Co and Pd have been reconstituted inside ApoF cages [199, 202, 203, 209, 210].

1.8.1 Magnetoferritin Synthesis

The fact that mineralisation sites are present within the confinement of the ferritin cavity can be used for size-controlled synthesis of cores whose magnetic properties can be tuned according to their size, as well as their composition. Size dependent magnetic property control has been demonstrated by changing the size of the protein cage or by changing the loading factor (i.e. the number of metal atoms per ferritin cage) inside the protein cavity, [211–213] and the composition of the core nanoparticle can be tuned by adding varying amounts of different metal salts during the mineralisation reaction [209, 214]. Common protocols for the synthesis of MF involve sequentially adding metal salts and hydrogen peroxide to ApoF solutions

[207, 215–218]. Synthesis is usually carried out at elevated temperatures (e.g. 65 °C) and under nitrogen atmosphere to achieve controlled oxidation of the metal salts and achieve mineralisation of a particular iron oxide phase (e.g. magnetite). The disassembly of ferritin subunits in aqueous solution at low pH (~pH 2) and subsequent re-assembly upon raising the pH is an alternative way to incorporate magnetic nanoparticles (4 and 6 nm) [219] and MRI contrast agents such as gadolinium-complexes [220] inside the protein shell. Loading of other functional molecules inside the ferritin cavity holds great promise for alternative delivery methods of anti-cancer drugs, [221, 222], photosensitisers that facilitate tumour treatment, [223] quantum dots [224] and radionuclides [225].

1.8.2 Applications of Magnetoferritin

The earliest application for MF was investigated in the field of MRI contrast agents. Early experiments showed that MF possessed a r2/r1 ratio one order of magnitude higher than similarly sized iron oxides synthesised in the absence of ferritin [215]. Since then, more detailed studies into tailoring the magnetic properties of MF contrast agents have been undertaken [213, 216, 226]. It has been shown that the r1 and r2 relaxivity of MF increased with increasing particle size. About 6 nm particles (which corresponds to about 5000 iron atoms per ferritin cage) had contrast enhancing properties comparable to those of commercially available ultra-small superparamagnetic iron oxide nanoparticle (USPION) contrast agents [213].

Furthermore, Uchida and co-workers have investigated the potential of using the MF protein shell to display peptides which target specific tissues. They genetically engineered ferritin cages to display RGD-4C which is known to bind to particular cell integrins (membrane receptors). Integrins are overexpressed in certain cancer cells (e.g. melanoma cells), [227] and loading such RGD-modified ferritins with a superparamagnetic core may be used to image cancerous tissue. Additionally, RGD-4C MF could also be used to image inflammatory events induced by macrophages (e.g. atherosclerotic plaque progression) because of the increased uptake of these modified magnetic protein nanoparticles by macrophages [226]. It is also possible to conjugate carbohydrates to the MF surface, which have a strong affinity to their corresponding carbohydrate binding lectins and may therefore be used as a probe in functional MRI [219]. A multimodal imaging platform can also be achieved by conjugating quantum dots to MF [228]. Finally, magnetoferritin can be used as a histological stain, because the iron oxide core catalyses the oxidation of peroxidase substrates, yielding a coloured product. Interestingly, tumour tissues can be identified with a great degree of specificity if a recombinant human heavy chain ferritin cage is used to produce magnetoferritin, because this particular protein cage binds to tumour cells that overexpress transferrin receptor 1 [229].

Apart from its use in the biomedical field, surface functionalised MF may also be used in environmental applications such as the decontamination of radioactive waste water. For example, a caesium specific chelating agent was attached to the

MF surface, and caesium-loaded MF could be sequestered from the aqueous solution by magnetic separation [230]. This principle could potentially be employed to remove other contaminants from waste water.

1.9 Chemical Protein Cationisation

Chemical cationisation of proteins has been shown to increase cellular delivery of otherwise poorly internalised proteins via adsorption-mediated endocytosis, and has therefore attracted much attention with regard to increasing delivery of protein based drugs and active molecules [231]. Protein cationisation is commonly achieved by carbodiimide-mediated coupling of diamines to negatively charged carboxyl groups on the protein surface [232]. To avoid crosslinking between proteins, an excess of diamine is added. Ethylenediamine is one of the most commonly used molecules for chemical protein cationisation, because its short chain length avoids steric bulk effects and hydrophobic interactions [233–235]. However, cationisation with ethylenediamine may lead to changes in native protein structure and consequent loss of biological activity, so that careful evaluation of the cationised product is recommended. Figure 1.20 shows a schematic of this cationisation reaction, which was also employed in this work to cationise MF.

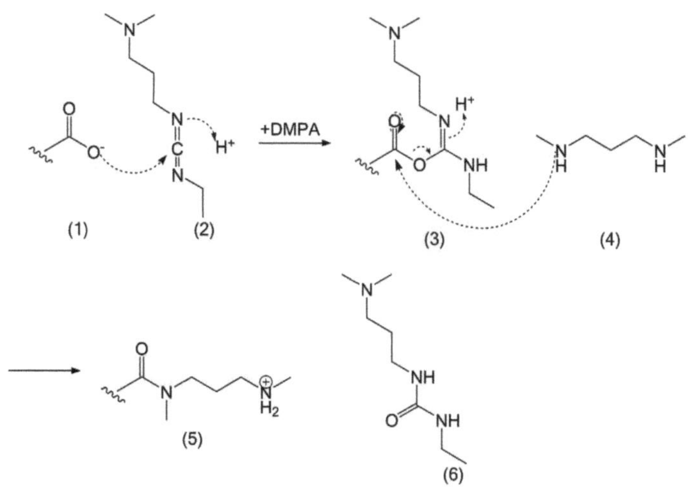

Fig. 1.20 Reaction schematic of protein cationisation used in this work. Carboxyl groups on the protein surface (*1*) react with the crosslinker 1-Ethyl-3-(3-dimethylaminopropyl)carbodiimide (EDC) (*2*) to form an O-acylisourea active ester intermediate (*3*). Nucleophile attack from an amine group of N,N-dimethyl-1,3-propanediamine (DMPA) (*4*) results in the formation of a peptide bond with the protein's carboxyl group (*5*) and release of an EDC by-product in the form of a soluble urea derivative (*6*). It should be noted that the O-acylisourea intermediate is unstable in aqueous solution and hydrolyses within a few minutes [236], leading to the regeneration of the carboxyl-group

1.9.1 Cationisation of Ferritin

Mammalian ferritins have a net negative charge at neutral pH, which is mainly due to the presence of aspartic and glutamic acid residues on the protein surface (see Fig. 1.21) [176]. Chemical cationisation of ferritin with DMPA has first been described in 1972 [237], with the aim to develop a TEM stain for cell membranes. In this early study, a variety of mammalian cells were fixed and sectioned for TEM, and sections were then "stained" with cationised ferritin. Using TEM, the authors found that cationised ferritin extensively adsorbed to the cell membrane, such that the cell surface could be visualised due to the electron-dense ferritin cores. Later, cationised ferritin was also used to develop a stain for anionic domains of bacterial cell membranes [238]. More recently, ferritin and other proteins were chemically cationised with DMPA to facilitate electrostatic attachment of surfactants to the cationised protein surface to produce solvent-free protein liquids [239].

1.10 Motivations for the Present Work

For magnetic cell separation, cell tracking with MRI and many other applications requiring the labelling of cells with magnetic material, sufficient cell uptake is crucial to achieve adequate cell magnetisation. At the same time, toxic effects due to the exposure to magnetic nanoparticles have to be avoided. Considering previous

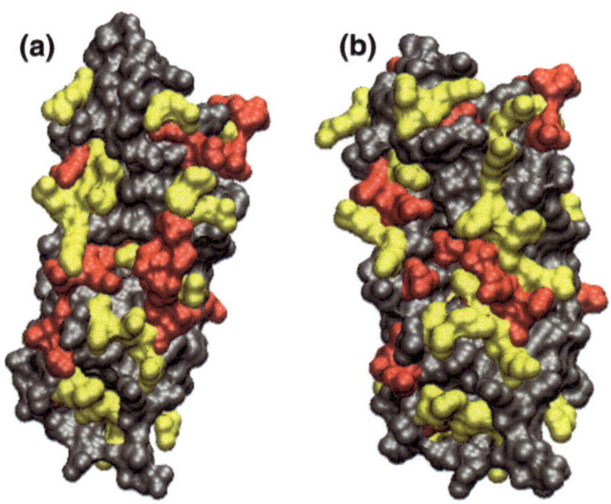

Fig. 1.21 Solvent-accessible surface area representations of mammalian ferritin subunits (1.4 Å probe radius). Acidic (*red*), basic (*yellow*) and neutral (*grey*) amino acid residues are depicted for **a** the light chain (PDB reference: 2W0O) and **b** the heavy chain (PDB reference: 2FHA). Number of cationisable sites (aspartate and glutamate residues): 27 on the light chain, 32 on the heavy chain. Horse spleen apoferritin used in this study consists to 90% of the light chain [240]

work discussed in Sect. 1.6.2, cationic charge-functionalisation has been identified as a strategy to achieve rapid and efficient uptake of SPIONs. Conventionally, labelling times have ranged from 1 to 24 h or more, but it is yet unknown if sufficient cell magnetisation can be achieved within shorter incubation times, possibly in the range of a few minutes. Therefore, the work presented here investigates the potential rapidity of magnetic labelling using cationised MF (cat-MF) and incubation times as short as one minute. Although a cationised particle surface has a great potential for rapid labelling, its toxic potential must be considered. Thus, the toxicity of cat-MF is also explored in this work. Finally, another advantage of charge-functionalisation is that efficient labelling is mediated by non-specific, electrostatic interactions. Therefore, efficient labelling may be extended to a variety of cell types, which is explored in this work using stem cells, and the bacteria *E. coli* and *S. aure*us.

1.10.1 Experimental Approaches and Rationales

The first step taken in this work was the synthesis of a novel cationised SPION derived from MF. A major advantage of MF is the ease with which its magnetic properties can be controlled, for example by tuning the size of the mineralised nanoparticle, [212, 213] or by doping it with metals such as cobalt [209, 214]. Furthermore, the surrounding protein shell affords aqueous solubility and a large canvas of addressable amino acids that can be chemically [219, 241] or genetically modified [227, 242, 243].

Both ferritin and MF have previously shown to be suitable for MACS [82, 241] and contrast enhancement in MRI [215, 244]. Ferritin has a lower magnetic moment compared to MF, [241] so using ferritin-labelled cells for MACS requires the application of high magnetic field strengths (4 T) [245]. To achieve these high fields, superconducting magnets need to be employed, which is and expensive and somewhat unpractical method in day-to-day lab work. MF has a higher magnetic moment, and thus commercially available permanent magnets can be used to capture MF-labelled cells. Antibody conjugation to MF has been explored and used to isolate specific cell types from a mixture of cell populations using a ferrograph, i.e. a device that uses a technique in which magnetically labelled particles or cells in suspension form a deposition pattern under a controlled flow and magnetic field [241]. Regarding contrast enhancement in MRI, MF has been shown to perform as well as commercially available ultra-small SPION-based contrast agents, [213] and functionalisation of the MF surface with targeting ligands has led to contrast generation in specific areas such as tumours or atherosclerotic plaques [226, 227].

Cationisation of the MF surface is a facile functionalisation route that has not yet been explored. It is to be expected that cationisation efficiencies are similar to those of ferritin, because the protein shell is not affected by synthetic mineralisation [205]. Given the ample evidence for an improvement in cell uptake efficiency of cationised nanoparticles and proteins discussed in Sects. 1.6.2 and 1.9, cell magnetisation and

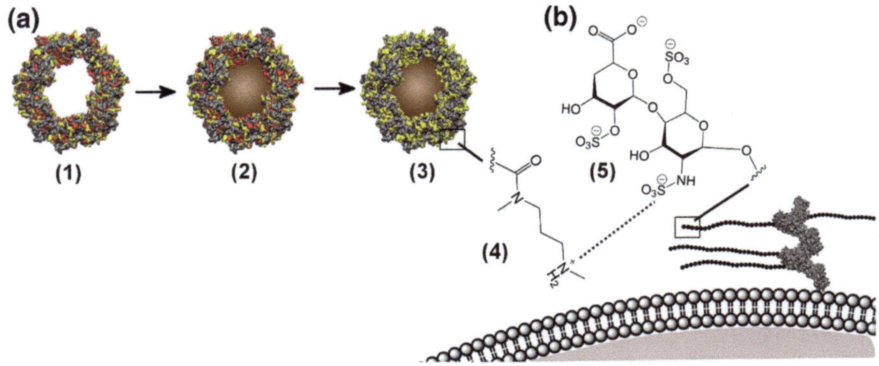

Fig. 1.22 Schematic of magnetoferritin synthesis, functionalisation and proposed labelling mechanism, not drawn to scale. **a** Apoferritin (*1*) is incubated with iron and cobalt salts to produce magnetoferritin (*2*). Carbodiimide-mediated crosslinking of *N,N'*-dimethyl-1,3-propanediamine (DMPA) to surface carboxyl groups yields cationised magnetoferritin (*3*). Solvent accessible surface area representations show the distribution of acidic (*red*) and basic (*yellow*) amino acid residues present on the protein surface. **b** Rapid cell labelling is mediated by electrostatic interactions (*dashed line*) between basic residues present at the cationised protein surface (*4*) and anionic groups on proteoglycans (e.g. heparan sulphate residues (*5*)) within the glycocalyx of the cytoplasmic membrane. Image re-printed with permission from Correia Carreira et al. [246]

iron content should be enhanced in cells exposed to cat-MF compared to native MF. The cationisation protocol developed here entails the conjugation of amine groups to the protein surface, which should result in a relatively lower positive surface charge compared to poly-cationic transfection agents. This may reduce any potential adverse effects that often result from high cationic charge densities, [147] but may still achieve rapid cell magnetisation using low extracellular iron concentrations via adsorption to anionic domains on the cell surface (e.g. proteoglycans). The basic concept underlying this work is summarised in Fig. 1.22.

1.11 Thesis Structure and Hypotheses Tested in This Work

The remaining chapters of this thesis describe the experimental methods used in this work (Chap. 2), followed by a summary of characterisation results obtained for MF and cat-MF (Chap. 3).

In Chap. 4, it was tested whether cationisation of MF had the capacity to rapidly and efficiently magnetise stem cells. As part of this hypothesis, effects of cat-MF concentration and exposure time were considered, and the persistence of magnetisation up to five weeks after labelling was assessed. For these experiments, magnetisation efficiency was evaluated using MACS and MRI. Additionally, it was also tested if rapid magnetic labelling was mediated through electrostatic adsorption of cat-MF onto anionic proteoglycans present on the cell surface.

In Chap. 5 it was tested whether cat-MF exposure induced adverse effects on mesenchymal stem cells, particularly with a view to the fact that cationic nanoparticles are known to be more toxic than their anionic or neutral counterparts [247, 248]. To test whether this was the case, a variety of toxicological profiling tools was employed, which assessed cell viability, membrane integrity, and cell proliferation and differentiation capacity.

In Chap. 6 it was tested if cat-MF could also be used to magnetise other cell types within short incubation times. Assuming that electrostatic interactions between the cationic particle surface and anionic domains on the cell surface mediate the rapidity and efficiency of the labelling process, it was hypothesised that bacteria can be equally efficiently magnetised, and subsequently removed or concentrated from water samples using MACS.

Finally, the global conclusion of this work and future opportunities will be summarised in Chap. 7.

References

1. Spaldin, N.A.: Magnetic Materials: Fundamentals and Applications. Cambridge University Press, Cambridge (2010)
2. Krishnan, K.M.: Fundamentals and Applications of Magnetic Materials. Oxford University Press (2016)
3. Jiles, D.C.: Introduction to Magnetism and Magnetic Materials, 1st edn. CRC Press (1998)
4. Du, Y.W., Xu, M.X., Wu, J., Shi, Y.B., Lu, H.X., Xue, R.H.: Magnetic properties of ultrafine nickel particles. J. Appl. Phys. **70**(10), 5903–5905 (1991)
5. Kodama, R.: Magnetic nanoparticles. J. Magn. Magn. Mater. **200**(1), 359–372 (1999)
6. Papaefthymiou, G.C.: Nanoparticle magnetism. Nano Today. **4**(5), 438–447 (2009)
7. Pankhurst, Q.A., Connolly, J., Jones, S.K., Dobson, J.: Applications of magnetic nanoparticles in biomedicine. J. Phys. D-Appl. Phys. **36**(13), R167–R181 (2003)
8. Bean, C., Livingston, J.: Superparamagnetism. J. Appl. Phys. **30**(4), S120–S129 (1959)
9. Palmberg, C., Dernis, H., Miguet, C.: Nanotechnology: an overview based on indicators and statistics (2009)
10. Grieve, K., Mulvaney, P., Grieser, F.: Synthesis and electronic properties of semiconductor nanoparticles/quantum dots. Curr. Opin. Colloid Interface Sci. **5**(1), 168–172 (2000)
11. Kelly, K.L., Coronado, E., Zhao, L.L., Schatz, G.C.: The optical properties of metal nanoparticles: the influence of size, shape, and dielectric environment. J. Phys. Chem. B **107** (3), 668–677 (2003)
12. Park, T.-J., Papaefthymiou, G.C., Viescas, A.J., Moodenbaugh, A.R., Wong, S.S.: Size-dependent magnetic properties of single-crystalline multiferroic Bifeo3 nanoparticles. Nano Lett. **7**(3), 766–772 (2007)
13. Arico, A.S., Bruce, P., Scrosati, B., Tarascon, J.-M., Van Schalkwijk, W.: Nanostructured materials for advanced energy conversion and storage devices. Nat. Mater. **4**(5), 366–377 (2005)
14. Sanvicens, N., Marco, M.P.: Multifunctional nanoparticles-properties and prospects for their use in human medicine. Trends Biotechnol. **26**(8), 425–433 (2008)

15. Oberdörster, G., Oberdörster, E., Oberdörster, J.: Nanotoxicology: an emerging discipline evolving from studies of ultrafine particles. Environ. Health Persp., 823–839 (2005)
16. Marradi, M., Martín-Lomas, M., PenadÉs, S.: Glyconanoparticles: polyvalent tools to study carbohydrate-based interactions. Adv. Carbohydr. Chem. Biochem. **64**, 211–290 (2010)
17. Sun, S., Zeng, H.: Size-controlled synthesis of magnetite nanoparticles. J. Am. Chem. Soc. **124**(28), 8204–8205 (2002)
18. Lee, S.-J., Jeong, J.-R., Shin, S.-C., Kim, J.-C., Kim, J.-D.: Synthesis and characterization of superparamagnetic maghemite nanoparticles prepared by coprecipitation technique. J. Magn. Magn. Mater. **282**, 147–150 (2004)
19. Johans, C., Pohjakallio, M., Ijäs, M., Ge, Y., Kontturi, K.: Control of particle size by pressure adjustment in cobalt nanoparticle synthesis. Colloids Surf., A **330**(1), 14–20 (2008)
20. Park, S.-J., Kim, S., Lee, S., Khim, Z.G., Char, K., Hyeon, T.: Synthesis and magnetic studies of uniform iron nanorods and nanospheres. J. Am. Chem. Soc. **122**(35), 8581–8582 (2000)
21. Park, J.-I., Cheon, J.: Synthesis of "Solid Solution" and "Core-Shell" type Cobalt-Platinum magnetic nanoparticles via transmetalation reactions. J. Am. Chem. Soc. **123**(24), 5743–5746 (2001)
22. Sun, S., et al.: Controlled Synthesis and Assembly of Fept Nanoparticles. J. Phys. Chem. B **107**(23), 5419–5425 (2003)
23. Lu, A.H., Salabas, E.L., Schüth, F.: Magnetic nanoparticles: synthesis, protection, functionalization, and application. Angew. Chemie. Int. Ed. **46**(8), 1222–1244 (2007)
24. Wu, L., Zhang, J., Watanabe, W.: Physical and chemical stability of drug nanoparticles. Adv. Drug Deliv. Rev. **63**(6), 456–469 (2011)
25. Bee, A., Massart, R., Neveu, S.: Synthesis of very fine maghemite particles. J. Magn. Magn. Mater. **149**(1), 6–9 (1995)
26. Sun, S., et al.: Monodisperse Mfe2o4 (M = Fe Co, Mn) Nanoparticles. J. Am. Chem. Soc. **126**(1), 273–279 (2004)
27. Reddy, L.H., Arias, J.L., Nicolas, J., Couvreur, P.: Magnetic nanoparticles: design and characterization, toxicity and biocompatibility, pharmaceutical and biomedical applications. Chem. Rev. **112**(11), 5818–5878 (2012)
28. Xie, J., Xu, C., Kohler, N., Hou, Y., Sun, S.: Controlled pegylation of monodisperse Fe3o4 nanoparticles for reduced non-specific uptake by macrophage cells. Adv. Mater. **19**(20), 3163–3166 (2007)
29. Veiseh, O., Gunn, J.W., Zhang, M.: Design and fabrication of magnetic nanoparticles for targeted drug delivery and imaging. Adv. Drug Deliv. Rev. **62**(3), 284–304 (2010)
30. Chekina, N., et al.: Fluorescent magnetic nanoparticles for biomedical applications. J. Mater. Chem. **21**(21), 7630–7639 (2011)
31. Lim, E.K., Huh, Y.M., Yang, J., Lee, K., Suh, J.S., Haam, S.: Ph-triggered drug-releasing magnetic nanoparticles for cancer therapy guided by molecular imaging by MRI. Adv. Mater. **23**(21), 2436–2442 (2011)
32. Dusinska, M., et al.: Towards an alternative testing strategy for nanomaterials used in nanomedicine: lessons from nanotest. Nanotoxicology **9**(sup1), 118–132 (2015)
33. Chen, J., et al.: Cationic nanoparticles induce nanoscale disruption in living cell plasma membranes. J. Phys. Chem. B **113**(32), 11179–11185 (2009)
34. Panessa-Warren, B., Warren, J., Wong, S., Misewich, J.: Biological cellular response to carbon nanoparticle toxicity. J. Phys.: Condens. Matter. **18**(33), S2185 (2006)
35. Marano, F., Hussain, S., Rodrigues-Lima, F., Baeza-Squiban, A., Boland, S.: Nanoparticles: molecular targets and cell signalling. Arch. Toxicol. **85**(7), 733–741 (2011)
36. Chen, M., von Mikecz, A.: Formation of nucleoplasmic protein aggregates impairs nuclear function in response to Sio 2 nanoparticles. Exp. Cell Res. **305**(1), 51–62 (2005)
37. Zuo, G., Huang, Q., Wei, G., Zhou, R., Fang, H.: Plugging into proteins: poisoning protein function by a hydrophobic nanoparticle. ACS Nano **4**(12), 7508–7514 (2010)
38. AshaRani, P., Low Kah Mun, G., Hande, M.P., Valiyaveettil, S.: Cytotoxicity and genotoxicity of silver nanoparticles in human cells. ACS Nano **3**(2), 279–290 (2008)

39. Klein, S., Sommer, A., Distel, L.V., Neuhuber, W., Kryschi, C.: Superparamagnetic iron oxide nanoparticles as radiosensitizer via enhanced reactive oxygen species formation. Biochem. Biophys. Res. Commun. **425**(2), 393–397 (2012)
40. Naqvi, S., et al.: Concentration-dependent toxicity of iron oxide nanoparticles mediated by increased oxidative stress. Int. J. Nanomed. **5**, 983 (2010)
41. Shander, A., Cappellini, M., Goodnough, L.: Iron overload and toxicity: the hidden risk of multiple blood transfusions. Vox Sang. **97**(3), 185–197 (2009)
42. Häfeli, U.O., et al.: Cell uptake and in vitro toxicity of magnetic nanoparticles suitable for drug delivery. Mol. Pharm. **6**(5), 1417–1428 (2009)
43. Arbab, A.S., et al.: Comparison of transfection agents in forming complexes with ferumoxides, cell labeling efficiency, and cellular viability. Mol. Imaging **3**(1), 24–32 (2004)
44. Berry, C.C., Wells, S., Charles, S., Aitchison, G., Curtis, A.S.: Cell response to Dextran-derivatised iron oxide nanoparticles post internalisation. Biomaterials **25**(23), 5405–5413 (2004)
45. Berry, C.C., Wells, S., Charles, S., Curtis, A.S.: Dextran and Albumin derivatised iron oxide nanoparticles: influence on fibroblasts in vitro. Biomaterials **24**(25), 4551–4557 (2003)
46. Daldrup-Link, H.E., et al.: Targeting of hematopoietic progenitor cells with Mr contrast agents. Radiology **228**, 760–767 (2003)
47. Singh, N., Jenkins, G.J., Asadi, R., Doak, S.H.: Potential toxicity of superparamagnetic iron oxide nanoparticles (Spion). Nano Rev **1**, (2010)
48. Zhu, M.-T., et al.: Endothelial dysfunction and inflammation induced by iron oxide nanoparticle exposure: risk factors for early atherosclerosis. Toxicol. Lett. **203**(2), 162–171 (2011)
49. Chithrani, B.D., Chan, W.C.: Elucidating the mechanism of cellular uptake and removal of protein-coated gold nanoparticles of different sizes and shapes. Nano Lett. **7**(6), 1542–1550 (2007)
50. Mahmoudi, M., Simchi, A., Milani, A., Stroeve, P.: Cell toxicity of superparamagnetic iron oxide nanoparticles. J. Colloid Interface Sci. **336**(2), 510–518 (2009)
51. Laurent, S., Burtea, C., Thirifays, C., Häfeli, U.O., Mahmoudi, M.: Crucial ignored parameters on nanotoxicology: the importance of toxicity assay modifications and "Cell Vision". PloS one **7**(1), e29997 (2012)
52. Miltenyi, S., Muller, W., Weichel, W., Radbruch, A.: High-gradient magnetic cell-separation with macs. Cytometry **11**(2), 231–238 (1990)
53. Mazuel, F., Reffay, M., Du, V., Bacri, J.C., Rieu, J.P., Wilhelm, C.: Magnetic flattening of stem-cell spheroids indicates a size-dependent elastocapillary transition. Phys. Rev. Lett. **114** (9) (2015)
54. Pan, Y., Du, X., Zhao, F., Xu, B.: Magnetic nanoparticles for the manipulation of proteins and cells. Chem. Soc. Rev. **41**(7), 2912–2942 (2012)
55. Dobson, J., Cartmell, S.H., Keramane, A., El Haj, A.J.: Principles and design of a novel magnetic force mechanical conditioning bioreactor for tissue engineering, stem cell conditioning, and dynamic in vitro screening. IEEE Trans. Nanobiosci. **5**(3), 173–177 (2006)
56. Kanczler, J.M., et al.: Controlled differentiation of human bone marrow stromal cells using magnetic nanoparticle technology. Tissue Eng. Part A **16**(10), 3241–3250 (2010)
57. Etoc, F., Lisse, D., Bellaiche, Y., Piehler, J., Coppey, M., Dahan, M.: Subcellular control of Rac-Gtpase signalling by magnetogenetic manipulation inside living cells. Nat. Nanotechnol. **8**(3), 193–198 (2013)
58. Kubo, T., Sugita, T., Shimose, S., Nitta, Y., Ikuta, Y., Murakami, T.: Targeted delivery of anticancer drugs with intravenously administered magnetic liposomes in Osteosarcoma-bearing hamsters. Int. J. Oncol. **17**(2), 309–324 (2000)
59. Souza, G.R., et al.: Three-dimensional tissue culture based on magnetic cell levitation. Nat. Nanotechnol. **5**(4), 291–296 (2010)

60. Jordan, A., Wust, P., Fähling, H., John, W., Hinz, A., Felix, R.: Inductive heating of ferrimagnetic particles and magnetic fluids: physical evaluation of their potential for hyperthermia. Int. J. Hyperth. **25**(7), 499–511 (2009)
61. Issels, R.: Hyperthermia combined with chemotherapy-biological rationale, clinical application, and treatment results. Oncol. Res. Treatment **22**(5), 374–381 (1999)
62. Bruners, P., et al.: Thermoablation of malignant kidney tumors using magnetic nanoparticles: an in vivo feasibility study in a rabbit model. Cardiovasc. Intervent. Radiol. **33**(1), 127–134 (2010)
63. Ferreira, V.M., et al.: Non-Contrast T1-Mapping detects acute myocardial edema with high diagnostic accuracy: a comparison to T2-weighted cardiovascular magnetic resonance. J. Cardiovasc. Magn. Reson. **14**(1), 42 (2012)
64. Guo, Y., et al.: Differentiation of clinically benign and malignant breast lesions using diffusion-weighted imaging. J. Magn. Reson. Imaging **16**(2), 172–178 (2002)
65. Geraldes, C.F.G.C., Laurent, S.: Classification and basic properties of contrast agents for magnetic resonance imaging. Contrast Media Mol. Imaging **4**(1), 1–23 (2009)
66. McRobbie, D.W., Moore, E.A., Graves, M.J., Prince, M.R.: Mri from Picture to Proton. Cambridge University Press (2006)
67. Weissleder, R., Elizondo, G., Wittenberg, J., Rabito, C.A., Bengele, H.H., Josephson, L.: Ultrasmall superparamagnetic iron-oxide—characterization of a new class of contrast agents for MR imaging. Radiology **175**(2), 489–493 (1990)
68. Michel, S.C.A., et al.: Preoperative breast cancer staging: MR imaging of the axilla with ultrasmall superparamagnetic iron oxide enhancement. Radiology **225**(2), 527–536 (2002)
69. Enochs, W.S., Harsh, G., Hochberg, F., Weissleder, R.: Improved delineation of human brain tumors on MR images using a long-circulating, superparamagnetic iron oxide agent. J. Magn. Reson. Imaging **9**(2), 228–232 (1999)
70. Rogers, W.J., Meyer, C.H., Kramer, C.M.: Technology insight: in vivo cell tracking by use of MRI. Nat. Clin. Pract. Cardiovasc. Med. **3**(10), 554–562 (2006)
71. Heyn, C., et al.: In vivo magnetic resonance imaging of single cells in mouse brain with optical validation. Magn. Reson. Med. **55**(1), 23–29 (2006)
72. Liberti, P.A., Rao, C.G., Terstappen, L.W.M.M.: Optimization of ferrofluids and protocols for the enrichment of breast tumor cells in blood. J. Magn. Magn. Mater. **225**(1–2), 301–307 (2001)
73. Paul, F., Melville, D., Roath, S., Warhurst, D.C.: A bench top magnetic separator for malarial parasite concentration. IEEE Trans. Magn. **17**(6), 2822–2824 (1981)
74. Kell, A.J., et al.: Vancomycin-modified nanoparticles for efficient targeting and preconcentration of Gram-Positive and Gram-Negative bacteria. ACS Nano **2**(9), 1777–1788 (2008)
75. Kemsheadl, J., Ugelstad, J.: Magnetic separation techniques: their application to medicine. Mol. Cell. Biochem. **67**(1), 11–18 (1985)
76. Kularatne, B.Y., Lorigan, P., Browne, S., Suvarna, S.K., Smith, N.O., Lawry, J.: Monitoring tumour cells in the peripheral blood of small cell lung cancer patients. Cytometry **50**(3), 160–167 (2002)
77. Zigeuner, R.E., Riesenberg, R., Pohla, H., Hofstetter, A., Oberneder, R.: Isolation of circulating cancer cells from whole blood by immunomagnetic cell enrichment and unenriched immunocytochemistry in vitro. J. Urol. **169**(2), 701–705 (2003)
78. Fong, C.Y., Peh, G.S., Gauthaman, K., Bongso, A.: Separation of Ssea-4 and Tra-1–60 labelled undifferentiated human embryonic stem cells from a heterogeneous cell population using magnetic-activated cell sorting (Macs) and Fluorescence-activated cell sorting (Facs). Stem Cell Rev. Rep. **5**(1), 72–80 (2009)
79. Morisada, S., Miyata, N., Iwahori, K.: Immunomagnetic separation of scum-forming bacteria using polyclonal antibody that recognizes mycolic acids. J. Microbiol. Methods **51** (2), 141–148 (2002)
80. Singh, S., Barick, K., Bahadur, D.: Surface engineered magnetic nanoparticles for removal of toxic metal ions and bacterial pathogens. J. Hazard. Mater. **192**(3), 1539–1547 (2011)

81. Huang, Y.-F., Wang, Y.-F., Yan, X.-P.: Amine-functionalized magnetic nanoparticles for rapid capture and removal of bacterial pathogens. Environ. Sci. Technol. **44**(20), 7908–7913 (2010)
82. Zborowski, M., Fuh, C.B., Green, R., Sun, L.P., Chalmers, J.J.: Analytical magnetapheresis of ferritin-labeled lymphocytes. Anal. Chem. **67**(20), 3702–3712 (1995)
83. Lanza, R., et al.: Essentials of Stem Cell Biology. Academic Press (2005)
84. Stocum, D.L.: Regenerative Biology and Medicine. Academic Press (2012)
85. Fuchs, E., Tumbar, T., Guasch, G.: Socializing with the neighbors: stem cells and their Niche. Cell **116**(6), 769–778 (2004)
86. Wu, J., Belmonte, J.C.I.: Stem cells: a renaissance in human biology research. Cell **165**(7), 1572–1585 (2016)
87. Tabar, V., Studer, L.: Pluripotent stem cells in regenerative medicine: challenges and recent progress. Nat. Rev. Genet. **15**(2), 82–92 (2014)
88. Weissman, I.L., Anderson, D.J., Gage, F.: Stem and progenitor cells: origins, phenotypes, lineage commitments, and transdifferentiations. Annu. Rev. Cell Dev. Biol. **17**(1), 387–403 (2001)
89. Forsberg, E.C., Bhattacharya, D., Weissman, I.L.: Hematopoietic stem cells. Stem Cell Rev. **2**(1), 23–30 (2006)
90. Caplan, A.I.: Mesenchymal stem cells. J. Orthop. Res. **9**(5), 641–650 (1991)
91. Ding, D.-C., Shyu, W.-C., Lin, S.-Z.: Mesenchymal stem cells. Cell Transplant. **20**(1), 5–14 (2011)
92. Dominici, M., et al.: Minimal criteria for defining multipotent mesenchymal stromal cells. The International Society for Cellular Therapy Position Statement. Cytotherapy **8**(4), 315–317 (2006)
93. Pittenger, M.F., et al.: Multilineage potential of adult human mesenchymal stem cells. Science **284**(5411), 143–147 (1999)
94. DiGirolamo, C.M., Stokes, D., Colter, D., Phinney, D.G., Class, R., Prockop, D.J.: Propagation and senescence of human marrow stromal cells in culture: a simple colony-forming assay identifies samples with the greatest potential to propagate and differentiate. Br. J. Haematol. **107**(2), 275–281 (1999)
95. Haynesworth, S., Goshima, J., Goldberg, V., Caplan, A.: Characterization of cells with osteogenic potential from human marrow. Bone **13**(1), 81–88 (1992)
96. Yoo, J.U., et al.: The chondrogenic potential of human bone-marrow-derived mesenchymal progenitor cells. J. Bone Joint Surg. **80**(12), 1745–1757 (1998)
97. Nakagami, H., Morishita, R., Maeda, K., Kikuchi, Y., Ogihara, T., Kaneda, Y.: Adipose tissue-derived stromal cells as a novel option for regenerative cell therapy. J. Atherosclerosis Thrombosis **13**(2), 77–81 (2006)
98. Wakitani, S., Saito, T., Caplan, A.I.: Myogenic cells derived from rat bone marrow mesenchymal stem cells exposed to 5-Azacytidine. Muscle Nerve **18**(12), 1417–1426 (1995)
99. Griffith, L.G., Naughton, G.: Tissue engineering-current challenges and expanding opportunities. Science **295**(5557), 1009–1014 (2002)
100. DiMarino, A.M., Caplan, A.I., Bonfield, T.L.: Mesenchymal stem cells in tissue repair. Front. Immunol. **4** (2013)
101. Nelson, T.J., Behfar, A., Yamada, S., Martinez-Fernandez, A., Terzic, A.: Stem cell platforms for regenerative medicine. Clin. Transl. Sci. **2**(3), 222–227 (2009)
102. Evans, M.J., Kaufman, M.H.: Establishment in Culture of Pluripotential Cells from Mouse Embryos. Nature **292**(5819), 154–156 (1981)
103. Martin, G.R.: Isolation of a pluripotent cell line from early mouse embryos cultured in medium conditioned by teratocarcinoma stem cells. Proc. Natl. Acad. Sci. **78**(12), 7634–7638 (1981)
104. Thomson, J.A., et al.: Embryonic stem cell lines derived from human blastocysts. Science **282**(5391), 1145–1147 (1998)
105. McLaren, A.: Ethical and social considerations of stem cell research. Nature **414**(6859), 129–131 (2001)

106. Nussbaum, J., et al.: Transplantation of undifferentiated murine embryonic stem cells in the heart: teratoma formation and immune response. FASEB J. **21**(7), 1345–1357 (2007)
107. Ankrum, J., Karp, J.M.: Mesenchymal stem cell therapy: two steps forward, one step back. Trends Mol. Med. **16**(5), 203–209 (2010)
108. Hayase, M., et al.: Committed neural progenitor cells derived from genetically modified bone marrow stromal cells ameliorate deficits in a rat model of stroke. J. Cereb. Blood Flow Metab. **29**(8), 1409–1420 (2009)
109. Segers, V.F.M., Lee, R.T.: Stem-cell therapy for cardiac disease. Nature **451**(7181), 937–942 (2008)
110. Venkataramana, N.K., et al.: Open-labeled study of unilateral autologous bone-marrow-derived mesenchymal stem cell transplantation in Parkinson's disease. Transl. Res. **155**(2), 62–70 (2010)
111. Petite, H., et al.: Tissue-engineered bone regeneration. Nat. Biotechnol. **18**(9), 959–963 (2000)
112. Wakitani, S., Imoto, K., Yamamoto, T., Saito, M., Murata, N., Yoneda, M.: Human autologous culture expanded bone marrow mesenchymal cell transplantation for repair of cartilage defects in osteoarthritic knees. Osteoarthritis Cartilage. **10**(3), 199–206 (2002)
113. Macchiarini, P., et al.: Clinical transplantation of a tissue-engineered airway. The Lancet **372** (9655), 2023–2030 (2008)
114. Ahrens, E.T., Flores, R., Xu, H., Morel, P.A.: In vivo imaging platform for tracking immunotherapeutic cells. Nat. Biotechnol. **23**(8), 983–987 (2005)
115. Hoehn, M., et al.: Monitoring of implanted stem cell migration in vivo: a highly resolved in vivo magnetic resonance imaging investigation of experimental stroke in rat. Proc. Natl. Acad. Sci. **99**(25), 16267–16272 (2002)
116. Chapon, C., Jackson, J.S., Aboagye, E.O., Herlihy, A.H., Jones, W.A., Bhakoo, K.K.: An in vivo multimodal imaging study using Mri and pet of stem cell transplantation after myocardial infarction in rats. Mol. Imag. Biol. **11**(1), 31–38 (2009)
117. Jing, X.H., et al.: In vivo Mr imaging tracking of magnetic iron oxide nanoparticle labeled, engineered, autologous bone marrow mesenchymal stem cells following intra-articular injection. Joint Bone Spine. **75**(4), 432–438 (2008)
118. Kyrtatos, P.G., et al.: Magnetic tagging increases delivery of circulating progenitors in vascular injury. JACC: Cardiovasc Intervent. **2**(8), 794–802 (2009)
119. Kirkham, G.R., et al.: Hyperpolarization of human mesenchymal stem cells in response to magnetic force. IEEE Trans. NanoBiosci. **9**(1), 71–74 (2010)
120. Cartmell, S.H., Dobson, J., Verschueren, S.B., El Haj, A.J.: Development of magnetic particle techniques for long-term culture of bone cells with intermittent mechanical activation. IEEE Trans. NanoBiosci. **1**(2), 92–97 (2002)
121. Hooper, L.V., Gordon, J.I.: Commensal host-bacterial relationships in the gut. Science **292** (5519), 1115–1118 (2001)
122. Guarner, F., Malagelada, J.-R.: Gut flora in health and disease. The Lancet **361**(9356), 512–519 (2003)
123. Cross, M.L.: Microbes versus microbes: immune signals generated by probiotic lactobacilli and their role in protection against microbial pathogens. FEMS Immunol. Med. Microbiol. **34**(4), 245–253 (2002)
124. Morens, D.M., Folkers, G.K., Fauci, A.S.: The challenge of emerging and re-emerging infectious diseases. Nature **430**(6996), 242–249 (2004)
125. Bartholomew, J.W., Mittwer, T.: The Gram Stain. Bacteriol. Rev. **16**(1), 1 (1952)
126. Madigan, M.T., Martinko, J.M., Bender, K.S.: Brock Biology of Microorganisms, 14th edn. Pearson Education (2014)
127. Weidenmaier, C., et al.: Lack of wall teichoic acids in staphylococcus aureus leads to reduced interactions with endothelial cells and to attenuated virulence in a Rabbit model of endocarditis. J. Infect. Dis. **191**(10), 1771–1777 (2005)
128. Akira, S., Uematsu, S., Takeuchi, O.: Pathogen recognition and innate immunity. Cell **124** (4), 783–801 (2006)

129. Markham, P.N., Neyfakh, A.A.: Efflux-mediated drug resistance in gram-positive bacteria. Curr. Opin. Microbiol. **4**(5), 509–514 (2001)
130. Costerton, J., Ingram, J., Cheng, K.: Structure and function of the cell envelope of gram-negative bacteria. Bacteriol. Rev. **38**(1), 87 (1974)
131. Beveridge, T.J.: Use of the gram stain in microbiology. Biotech. Histochem. **76**(3), 111–118 (2001)
132. Vaara, M.: Antibiotic-Supersusceptible Mutants of Escherichia coli and Salmonella typhimurium. Antimicrob Agents Chemother. **37**(11), 2255 (1993)
133. Nikaido, H.: Molecular basis of bacterial outer membrane permeability revisited. Microbiol. Mol. Biol. Rev. **67**(4), 593–656 (2003)
134. Nikaido, H.: Multidrug Efflux Pumps of Gram-Negative Bacteria. J. Bacteriol. **178**(20), 5853 (1996)
135. Aderem, A., Underhill, D.M.: Mechanisms of phagocytosis in macrophages. Annu. Rev. Immunol. **17**(1), 593–623 (1999)
136. Conner, S.D., Schmid, S.L.: Regulated portals of entry into the cell. Nature **422**(6927), 37–44 (2003)
137. Zhao, F., Zhao, Y., Liu, Y., Chang, X., Chen, C., Zhao, Y.: Cellular uptake, intracellular trafficking, and cytotoxicity of nanomaterials. Small **7**(10), 1322–1337 (2011)
138. Dausend, J., et al.: Uptake mechanism of oppositely charged fluorescent nanoparticles in hela cells. Macromol. Biosci. **8**(12), 1135–1143 (2008)
139. Jiang, X., Musyanovych, A., Röcker, C., Landfester, K., Mailänder, V., Nienhaus, G.U.: Specific effects of surface carboxyl groups on anionic polystyrene particles in their interactions with mesenchymal stem cells. Nanoscale **3**(5), 2028–2035 (2011)
140. Chaudhary, S., et al.: Elucidating the function of penetratin and a static magnetic field in cellular uptake of magnetic nanoparticles. Pharmaceuticals. **6**(2), 204–222 (2013)
141. Schafer, D.A.: Coupling actin dynamics and membrane dynamics during endocytosis. Curr. Opin. Cell Biol. **14**(1), 76–81 (2002)
142. Swanson, J.A., Watts, C.: Macropinocytosis. Trends Cell Biol. **5**(11), 424–428 (1995)
143. Iversen, T.-G., Skotland, T., Sandvig, K.: Endocytosis and intracellular transport of nanoparticles: present knowledge and need for future studies. Nano Today **6**(2), 176–185 (2011)
144. Levy, M., et al.: Long term in vivo biotransformation of iron oxide nanoparticles. Biomaterials **32**(16), 3988–3999 (2011)
145. Wilhelm, C., Gazeau, F.: Universal cell labelling with anionic magnetic nanoparticles. Biomaterials **29**(22), 3161–3174 (2008)
146. Kolosnjaj-Tabi, J., et al.: The one year fate of iron oxide coated gold nanoparticles in mice. ACS Nano (2015)
147. Nel, A.E., et al.: Understanding biophysicochemical interactions at the nano-bio interface. Nat. Mater. **8**(7), 543–557 (2009)
148. Decuzzi, P., Ferrari, M.: The receptor-mediated endocytosis of nonspherical particles. Biophys. J. **94**(10), 3790–3797 (2008)
149. Gao, H., Shi, W., Freund, L.B.: Mechanics of receptor-mediated endocytosis. Proc. Natl. Acad. Sci. USA **102**(27), 9469–9474 (2005)
150. Mailander, V., et al.: Carboxylated superparamagnetic iron oxide particles label cells intracellularly without transfection agents. Mol. Imag. Biol. **10**(3), 138–146 (2008)
151. Dodd, S.J., Williams, M., Suhan, J.P., Williams, D.S., Koretsky, A.P., Ho, C.: Detection of single mammalian cells by high-resolution magnetic resonance imaging. Biophys. J. **76**(1 Pt 1), 103–109 (1999)
152. Smith, C.-A.M., de la Fuente, J., Pelaz, B., Furlani, E.P., Mullin, M., Berry, C.C.: The effect of static magnetic fields and tat peptides on cellular and nuclear uptake of magnetic nanoparticles. Biomaterials **31**(15), 4392–4400 (2010)
153. Barnes, A.L., Wassel, R.A., Mondalek, F., Chen, K., Dormer, K.J., Kopke, R.D.: Magnetic characterization of superparamagnetic nanoparticles pulled through model membranes. Biomagn. Res. Technol. **5**(1), 1 (2007)

154. Ahrens, E.T., Feili-Hariri, M., Xu, H., Genove, G., Morel, P.A.: Receptor-mediated endocytosis of iron-oxide particles provides efficient labeling of dendritic cells for in vivo Mr imaging. Magn. Reson. Med. **49**(6), 1006–1013 (2003)
155. Bulte, J.W.M., et al.: Magnetodendrimers allow endosomal magnetic labeling and in vivo tracking of stem cells. Nat. Biotechnol. **19**(12), 1141–1147 (2001)
156. Frank, J.A., et al.: Clinically applicable labeling of mammalian and stem cells by combining superparamagnetic iron oxides and transfection agents (vol 228, Pg 480, 2003). Radiology **229**(2), 610 (2003)
157. Schwarz, S., et al.: Synthetic and biogenic magnetite nanoparticles for tracking of stem cells and dendritic cells. J. Magn. Magn. Mater. **321**(10), 1533–1538 (2009)
158. Josephson, L., Tung, C.H., Moore, A., Weissleder, R.: High-efficiency intracellular magnetic labeling with novel superparamagnetic-tat peptide conjugates. Bioconjug. Chem. **10**(2), 186–191 (1999)
159. Lewin, M., et al.: Tat peptide-derivatized magnetic nanoparticles allow in vivo tracking and recovery of progenitor cells. Nat. Biotechnol. **18**(4), 410–414 (2000)
160. Suh, J.S., et al.: Efficient labeling of mesenchymal stem cells using cell permeable magnetic nanoparticles. Biochem. Biophys. Res. Commun. **379**(3), 669–675 (2009)
161. Dejardin, T., et al.: Influence of both a static magnetic field and penetratin on magnetic nanoparticle delivery into fibroblasts. Nanomedicine **6**(10), 1719–1731 (2011)
162. Montet-Abou, K., Montet, X., Weissleder, R., Josephson, L.: Cell internalization of magnetic nanoparticles using transfection agents. Mol. Imaging. **6**(1), 1–9 (2007)
163. Arbab, A.S., et al.: Efficient magnetic cell labeling with protamine sulfate complexed to ferumoxides for cellular Mri. Blood **104**(4), 1217–1223 (2004)
164. Wilhelm, C., et al.: Magnetic control of vascular network formation with magnetically labeled endothelial progenitor cells. Biomaterials **28**(26), 3797–3806 (2007)
165. Babic, M., et al.: Poly (N, N-Dimethylacrylamide)-coated maghemite nanoparticles for stem cell labeling. Bioconjug. Chem. **20**(2), 283–294 (2009)
166. Guo, R.M., et al.: Controllable labelling of stem cells with a novel superparamagnetic iron oxide-loaded cationic nanovesicle for Mr imaging. Eur. Radiol. **22**(11), 2328–2337 (2012)
167. Lu, C.-W., et al.: Bifunctional magnetic silica nanoparticles for highly efficient human stem cell labeling. Nano Lett. **7**(1), 149–154 (2007)
168. Wilhelm, C., Billotey, C., Roger, J., Pons, J., Bacri, J.-C., Gazeau, F.: Intracellular uptake of anionic superparamagnetic nanoparticles as a function of their surface coating. Biomaterials **24**(6), 1001–1011 (2003)
169. Arnold, L., Dagan, A., Gutheil, J., Kaplan, N.: Antineoplastic activity of Poly (L-Lysine) with some ascites tumor cells. Proc. Natl. Acad. Sci. **76**(7), 3246–3250 (1979)
170. Bull, B., Huse, W., Brauer, F., Korpman, R.: Heparin therapy during extracorporeal circulation. Ii. The use of a dose-response curve to individualize heparin and protamine dosage. J. Thoracic Cardiovasc. Surg. **69**(5), 685–689 (1975)
171. Zhang, D., et al.: Functionalization of whole-cell bacterial reporters with magnetic nanoparticles. Microb. Biotechnol. **4**(1), 89–97 (2011)
172. Gu, H.W., Xu, K.M., Xu, C.J., Xu, B.: Biofunctional magnetic nanoparticles for protein separation and pathogen detection. Chem. Commun. **9**, 941–949 (2006)
173. Gu, H., Ho, P.-L., Tsang, K.W., Wang, L., Xu, B.: Using biofunctional magnetic nanoparticles to capture vancomycin-resistant Enterococci and other gram-positive bacteria at ultralow concentration. J. Am. Chem. Soc. **125**(51), 15702–15703 (2003)
174. Gao, J., Li, L., Ho, P.L., Mak, G.C., Gu, H., Xu, B.: Combining fluorescent probes and biofunctional magnetic nanoparticles for rapid detection of bacteria in human blood. Adv. Mater. **18**(23), 3145–3148 (2006)
175. Lin, Y.-S., Tsai, P.-J., Weng, M.-F., Chen, Y.-C.: Affinity capture using vancomycin-bound magnetic nanoparticles for the Maldi-Ms analysis of bacteria. Anal. Chem. **77**(6), 1753–1760 (2005)
176. Harrison, P.M., Arosio, P.: The ferritins: molecular properties, iron storage function and cellular regulation. Biochim. Biophys. Acta **1275**(3), 161–203 (1996)

177. Theil, E.C., Behera, R.K., Tosha, T.: Ferritins for chemistry and for life. Coord. Chem. Rev. **257**(2), 579–586 (2013)
178. Uchida, M., Kang, S., Reichhardt, C., Harlen, K., Douglas, T.: The ferritin superfamily: supramolecular templates for materials synthesis. Biochim. Biophys. Acta **1800**(8), 834–845 (2010)
179. Gutteridge, J.: Iron and oxygen: a biologically damaging mixture. Acta paediatrica Scandinavica. Supplement. **361**, 78–85 (1988)
180. Chasteen, N.D., Harrison, P.M.: Mineralization in ferritin: an efficient means of iron storage. J. Struct. Biol. **126**(3), 182–194 (1999)
181. Theil, E.C.: Ferritin protein nanocages use ion channels, catalytic sites, and nucleation channels to manage iron/oxygen chemistry. Curr. Opin. Chem. Biol. **15**(2), 304–311 (2011)
182. Yamashita, I., Iwahori, K., Kumagai, S.: Ferritin in the Field of Nanodevices. Biochimica et Biophysica Acta (BBA)-General Sub. **1800,** (8), 846–857 (2010)
183. Arosio, P., Adelman, T.G., Drysdale, J.W.: On ferritin heterogeneity. Further evidence for heteropolymers. J. Biol. Chem. **253**(12), 4451–4458 (1978)
184. Bomford, A., Conlon-Hollingshead, C., Munro, H.: Adaptive responses of rat tissue isoferritins to iron administration. changes in subunit synthesis, isoferritin abundance, and capacity for iron storage. J. Biol. Chem. **256**(2), 948–955 (1981)
185. Linder, M.C., Kakavandi, H.R., Miller, P., Wirth, P.L., Nagel, G.M.: Dissociation of ferritins. Arch. Biochem. Biophys. **269**(2), 485–496 (1989)
186. Theil, E.C., Takagi, H., Small, G.W., He, L., Tipton, A., Danger, D.: The ferritin iron entry and exit problem. Inorg. Chim. Acta **297**(1), 242–251 (2000)
187. Tosha, T., Ng, H.-L., Bhattasali, O., Alber, T., Theil, E.C.: Moving metal ions through ferritin−protein nanocages from three-fold pores to catalytic sites. J. Am. Chem. Soc. **132** (41), 14562–14569 (2010)
188. Haldar, S., Bevers, L.E., Tosha, T., Theil, E.C.: Moving iron through ferritin protein nanocages depends on residues throughout each four A-Helix bundle subunit. J. Biol. Chem. **286**(29), 25620–25627 (2011)
189. Turano, P., Lalli, D., Felli, I.C., Theil, E.C., Bertini, I.: Nmr reveals pathway for ferric mineral precursors to the central cavity of ferritin. Proc. Natl. Acad. Sci. **107**(2), 545–550 (2010)
190. Pierre, T.G.S., et al.: Organ-specific crystalline structures of ferritin cores Inβ-Thalassemia/Hemoglobin E. Biol. Metals. **4**(3), 162–165 (1991)
191. Takeda, S., Yoshimura, H., Endo, S., Takahashi, T., Nagayama, K.: Control of crystal forms of apoferritin by site-directed mutagenesis. Proteins: Struct., Funct., Bioinf. **23**(4), 548–556 (1995)
192. Mayer, D.E., Rohrer, J.S., Schoeller, D.A., Harris, D.C.: Fate of oxygen during ferritin iron incorporation. Biochemistry **22**(4), 876–880 (1983)
193. Linder, M.C.: Nutritional Biochemistry and Metabolism: With Clinical Applications (1991)
194. Aisen, P., Listowsky, I.: Iron transport and storage proteins. Annu. Rev. Biochem. **49**(1), 357–393 (1980)
195. Harford, J., Casey, J., Koeller, D., Klausner, R.: Structure, function and regulation of the transferrin receptor: insights from molecular biology. Intracellular Trafficking of Proteins, pp. 302–334. Cambridge University Press, New York (1991)
196. Harford, J.B., Klausner, R.D.: Coordinate post-transcriptional regulation of ferritin and transferrin receptor expression: the role of regulated Rna-protein interaction. In Post-Transcriptional Control of Gene Expression, pp 399–409. Springer (1990)
197. Pawelczyk, E., Arbab, A.S., Pandit, S., Hu, E., Frank, J.A.: Expression of transferrin receptor and ferritin following ferumoxides-protamine sulfate labeling of cells: implications for cellular magnetic resonance imaging. NMR Biomed. **19**(5), 581–592 (2006)
198. Kidane, T.Z., Sauble, E., Linder, M.C.: Release of iron from ferritin requires lysosomal activity. Am. J. Physiol.-Cell Physiol. **291**(3), C445–C455 (2006)

199. Clemente-León, M., Coronado, E., Soriano-Portillo, A., Gálvez, N., Domínguez-Vera, J.M.: Permanent magnetism in apoferritin-encapsulated Pd nanoparticles. J. Mater. Chem. **17**(1), 49–51 (2007)
200. Ueno, T., et al.: Process of accumulation of metal ions on the interior surface of apo-ferritin: crystal structures of a series of apo-ferritins containing variable quantities of Pd(Ii) ions. J. Am. Chem. Soc. **131**(14), 5094–5100 (2009)
201. Gálvez, N., Fernandez, B., Valero, E., Sánchez, P., Cuesta, R., Domínguez-Vera, J.M.: Apoferritin as a nanoreactor for preparing metallic nanoparticles. C. R. Chim. **11**(10), 1207–1212 (2008)
202. Gálvez, N., Sánchez, P., Domínguez-Vera, J.M., Soriano-Portillo, A., Clemente-León, M., Coronado, E.: Apoferritin-encapsulated Ni and Co superparamagnetic nanoparticles. J. Mater. Chem. **16**(26), 2757–2761 (2006)
203. Okuda, M., Iwahori, K., Yamashita, I., Yoshimura, H.: Fabrication of nickel and chromium nanoparticles using the protein cage of apoferritin. Biotechnol. Bioeng. **84**(2), 187–194 (2003)
204. Mackle, P., Charnock, J.M., Garner, C.D., Meldrum, F.C., Mann, S.: Characterization of the manganese core of reconstituted ferritin by X-ray absorption spectroscopy. J. Am. Chem. Soc. **115**(18), 8471–8472 (1993)
205. Douglas, T., Stark, V.T.: Nanophase Cobalt Oxyhydroxide mineral synthesized within the protein cage of ferritin. Inorg. Chem. **39**(8), 1828–1830 (2000)
206. Meldrum, F.C., Douglas, T., Levi, S., Arosio, P., Mann, S.: Reconstitution of manganese oxide cores in horse spleen and recombinant ferritins. J. Inorg. Biochem. **58**(1), 59–68 (1995)
207. Meldrum, F.C., Heywood, B.R., Mann, S.: Magnetoferritin. In vitro synthesis of a novel magnetic protein. Science **257**(5069), 522–523 (1992)
208. Pankhurst, Q., Betteridge, S., Dickson, D., Douglas, T., Mann, S., Frankel, R.B.: Mössbauer spectroscopic and magnetic studies of magnetoferritin. Hyperfine Interact. **91**(1), 847–851 (1994)
209. Klem, M.T., Resnick, D.A., Gilmore, K., Young, M., Idzerda, Y.U., Douglas, T.: Synthetic control over magnetic moment and exchange bias in all-oxide materials encapsulated within a spherical protein cage. J. Am. Chem. Soc. **129**(1), 197–201 (2007)
210. Warne, B., Kasyutich, O., Mayes, E.L., Wiggins, J.A., Wong, K.K.: Self Assembled nanoparticulate Co: Pt for data storage applications. IEEE Trans. Magn. **36**(5), 3009–3011 (2000)
211. Gilmore, K., Idzerda, Y.U., Klem, M.T., Allen, M., Douglas, T., Young, M.: Surface contribution to the anisotropy energy of spherical magnetite particles. J. Appl. Phys. **97**(10) (2005)
212. Martinez-Perez, M., et al.: Size-dependent properties of magnetoferritin. Nanotechnology **21** (46), 465707 (2010)
213. Uchida, M., et al.: A human ferritin iron oxide nano-composite magnetic resonance contrast agent. Magn. Reson. Med. **60**(5), 1073–1081 (2008)
214. Okuda, M., Eloi, J.C., Sarua, A., Jones, S.E.W., Schwarzacher, W.: Energy barrier distribution for dispersed mixed oxide magnetic nanoparticles. J. Appl. Phys. **111**(7) (2012)
215. Bulte, J.W.M., et al.: Magnetoferritin—Characterization of a novel superparamagnetic MR contrast agent. JMRI-J. Magn. Resonan. Imaging **4**(3), 497–505 (1994)
216. Clavijo Jordan, V., Caplan, M.R., Bennett, K.M.: Simplified synthesis and relaxometry of magnetoferritin for Magnetic Resonance Imaging. Magn. Resonan. Med. **64**(5), 1260–1266 (2010)
217. Okuda, M., Eloi, J.-C., Jones, S.E.W., Sarua, A., Richardson, R.M., Schwarzacher, W.: Fe3o4 nanoparticles: protein-mediated crystalline magnetic superstructures. Nanotechnology **23**(41), 415601 (2012)
218. Wong, K.K., Douglas, T., Gider, S., Awschalom, D.D., Mann, S.: Biomimetic synthesis and characterization of magnetic proteins (Magnetoferritin). Chem. Mater. **10**(1), 279–285 (1998)

219. Valero, E., et al.: Magnetic nanoparticles-templated assembly of protein subunits: a new platform for carbohydrate-based MRI nanoprobes. J. Am. Chem. Soc. **133**(13), 4889–4895 (2011)
220. Aime, S., Frullano, L., Geninatti Crich, S.: Compartmentalization of a gadolinium complex in the apoferritin cavity: a route to obtain high relaxivity contrast agents for Magnetic Resonance Imaging. Angew. Chem. Int. Ed. Engl. **41**(6), 1017–1019 (2002)
221. Liang, M., et al.: H-Ferritin–Nanocaged doxorubicin nanoparticles specifically target and kill tumors with a single-dose injection. Proc. Natl. Acad. Sci. **111**(41), 14900–14905 (2014)
222. Xing, R.M., et al.: Characterization and cellular uptake of platinum anticancer drugs encapsulated in apoferritin. J. Inorg. Biochem. **103**(7), 1039–1044 (2009)
223. Zhen, Z., et al.: Ferritin nanocages to encapsulate and deliver photosensitizers for efficient photodynamic therapy against cancer. ACS Nano **7**(8), 6988–6996 (2013)
224. Wong, K.K., Mann, S.: Biomimetic synthesis of cadmium sulfide-ferritin nanocomposites. Adv. Mater. **8**(11), 928–932 (1996)
225. Lin, X., et al.: Chimeric ferritin nanocages for multiple function loading and multimodal imaging. Nano Lett. **11**(2), 814–819 (2011)
226. Terashima, M., et al.: Human ferritin cages for imaging vascular macrophages. Biomaterials **32**(5), 1430–1437 (2011)
227. Uchida, M., et al.: Targeting of cancer cells with ferrimagnetic ferritin cage nanoparticles. J. Am. Chem. Soc. **128**(51), 16626–16633 (2006)
228. Fernandez, B., Galvez, N., Cuesta, R., Hungria, A.B., Calvino, J.J., Domínguez-Vera, J.M.: Quantum dots decorated with magnetic bionanoparticles. Adv. Func. Mater. **18**(24), 3931–3935 (2008)
229. Fan, K., et al.: Magnetoferritin nanoparticles for targeting and visualizing tumour tissues. Nat. Nanotechnol. **7**(7), 459–464 (2012)
230. Urban, I., Ratcliffe, N.M., Duffield, J.R., Elder, G.R., Patton, D.: Functionalized paramagnetic nanoparticles for waste water treatment. Chem. Commun. (Camb.) **46**(25), 4583–4585 (2010)
231. Futami, J., Kitazoe, M., Murata, H., Yamada, H.: Exploiting Protein Cationization Techniques in Future Drug Development (2007)
232. Yamada, H., Imoto, T., Fujita, K., Okazaki, K., Motomura, M.: Selective modification of Aspartic Acid-101 in lysozyme by carbodiimide reaction. Biochemistry **20**(17), 4836–4842 (1981)
233. Futami, J., et al.: Preparation of potent cytotoxic ribonucleases by cationization: enhanced cellular uptake and decreased interaction with ribonuclease inhibitor by chemical modification of carboxyl groups. Biochemistry **40**(25), 7518–7524 (2001)
234. Griffin, D.E., Giffels, J.: Study of protein characteristics that influence entry into the cerebrospinal fluid of normal mice and mice with encephalitis. J. Clin. Investig. **70**(2), 289 (1982)
235. Kumagai, A., Eisenberg, J.B., Pardridge, W.: Absorptive-mediated endocytosis of cationized albumin and a beta-endorphin-cationized albumin chimeric peptide by isolated brain capillaries. Model system of blood-brain barrier transport. J. Biol. Chem. **262**(31), 15214–15219 (1987)
236. Wrobel, N., Schinkinger, M., Mirsky, V.M.: A novel ultraviolet assay for testing side reactions of carbodiimides. Anal. Biochem. **305**(2), 135–138 (2002)
237. Danon, D., Skutelsk, E., Marikovs, Y., Goldstei, L.: Use of cationized ferritin as a label of negative charges on cell surfaces. J. Ultrastruct. Res. **38**(5–6), 500–510 (1972)
238. Anderson, K.L.: Cationized ferritin as a stain for electron microscopic observation of bacterial ultrastructure. Biotechnol. Histochem. **73**(5), 278–288 (1998)
239. Perriman, A.W., Cölfen, H., Hughes, R.W., Barrie, C.L., Mann, S.: Solvent-free protein liquids and liquid crystals. Angew. Chem. Int. Ed. **48**(34), 6242–6246 (2009)
240. Stefanini, S., Chiancone, E., Arosio, P., Finazzi-Agro, A., Antonini, E.: Structural heterogeneity and subunit composition of horse ferritins. Biochemistry **21**(10), 2293–2299 (1982)

241. Zborowski, M., et al.: Immunomagnetic isolation of magnetoferritin-labeled cells in a modified ferrograph. Cytometry **24**(3), 251–259 (1996)
242. Ikezoe, Y., et al.: Growth of giant two-dimensional crystal of protein molecules from a three-phase contact line. Langmuir **24**(22), 12836–12841 (2008)
243. Yamashita, K., et al.: Selective nanoscale positioning of ferritin and nanoparticles by means of target-specific peptides. Small **2**(10), 1148–1152 (2006)
244. Fukunaga, M., et al.: Layer-specific variation of iron content in cerebral cortex as a source of MRI contrast. Proc. Natl. Acad. Sci. USA **107**(8), 3834–3839 (2010)
245. Owen, C.S., Lindsay, J.G.: Ferritin as a label for high-gradient magnetic separation. Biophys. J. **42**(2), 145–150 (1983)
246. Carreira, S.C., Armstrong, J., Seddon, A., Perriman, A., Hartley-Davies, R., Schwarzacher, W.: Ultra-fast stem cell labelling using cationised magnetoferritin. Nanoscale **8**(14), 7474–7483 (2016)
247. Frohlich, E.: The role of surface charge in cellular uptake and cytotoxicity of medical nanoparticles. Int. J. Nanomed. **7**, 5577–5591 (2012)
248. Goodman, C.M., McCusker, C.D., Yilmaz, T., Rotello, V.M.: Toxicity of gold nanoparticles functionalized with cationic and anionic side chains. Bioconjug. Chem. **15**(4), 897–900 (2004)

Chapter 2
Materials and Methods

This chapter contains the theory and protocols of experimental techniques used in this work, as well as details on the materials and methods for each experiment that was performed. First, general materials and methods are summarised, followed by a description of more specialised methods such as chromatography techniques and nanoparticle characterisation methods. The protocol for MF synthesis and cationisation is included, as well as basic protocols for stem cell culture. After that, experiments carried out to investigate stem cell labelling with cat-MF are described, followed by the methods that were used to assess cat-MF toxicity. Finally, bacterial culture protocols are described, as well as the experiments conducted to investigate cat-MF labelling of bacteria.

2.1 General Materials

Unless otherwise stated, chemicals were purchased from Sigma Aldrich (UK) and used as received. Deionised water (dH_2O) was obtained from a Merck Millipore ultrapure water system, with a resistivity of approximately 18 MΩ cm.

2.2 General Methods

2.2.1 *Buffer Preparation*

Theory
Buffer solutions prevent a rapid change in pH when acids or bases are added, such that the pH of a solution can be maintained at a near constant value. This is important in, for example, biochemical applications because many proteins and

S. Correia Carreira, *Rapid Cell Magnetisation Using Cationised Magnetoferritin*,
Springer Theses, DOI 10.1007/978-3-319-60333-9_2

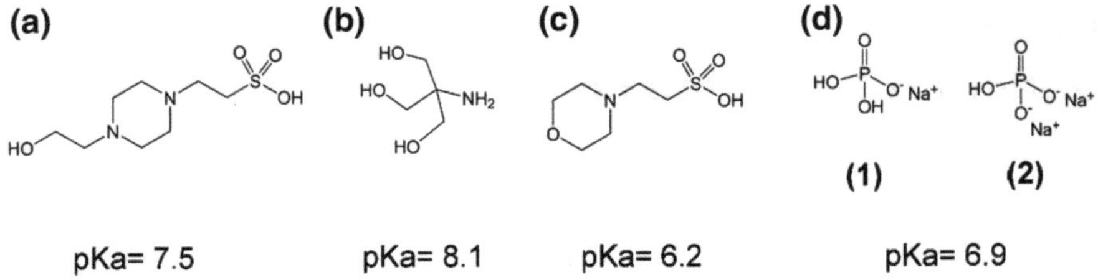

Fig. 2.1 Chemical formulae of buffering agents and their pK_a values at 25 °C. **a** HEPES, **b** Tris, **c** MES, and **d** phosphate buffer components: monosodium phosphate (*1*) and its conjugated base disodium phosphate (*2*)

enzymes only retain their structure and function over a narrow pH range [1]. There are a variety of buffering agents, such as Tris, HEPES and MES that have different buffering ranges. Mixtures of acids and their conjugate base can also be used to make a buffer solutions, as is the case with phosphate buffers (Fig. 2.1).

Buffer solutions resist pH change because of the equilibrium between the acid (HA) and conjugate base (A^-) form of the buffering agent or the acid-base pair.

$$HA \leftrightharpoons H^+ + A^-$$

When acid is added, the additional protons are "captured" by the conjugate base form of the buffer agent and the equilibrium shifts towards the left hand side. The opposite is true when base is added. At a certain pH value, the concentration of acid and conjugate base is equal. This is described by the pK_a value. Thus, the pH changes relatively slowly in the buffer region of pH = pKa ± 1.

Method

Buffer salts were dissolved in 80–90% of the final volume of dH_2O, the pH was adjusted with dilute hydrochloric acid (HCl) or sodium hydroxide solutions (NaOH) to the desired value, and the buffer solution made up to the final volume with dH_2O. Stock solutions were prepared for all buffers, which were diluted to the desired concentrations as required for individual experiments.

HEPES buffer was prepared as a 500 mM stock solution using 119 g of 4-(2-hydroxyethyl)-1-piperazineethanesulfonic acid (HEPES) in one litre dH_2O and the pH adjusted to 8.6 with NaOH.

Tris buffer was prepared as a 500 mM stock solution by dissolving 60.5 g of tris (hydroxymethyl)aminomethane (Tris) in one litre dH_2O, and the pH adjusted to pH 8.0, 7.6, or 7.0 using HCl.

Phosphate buffer (pH 7) was prepared from 57.2 g monosodium phosphate and 74.2 g disodium phosphate in one litre dH_2O to yield a 1 M stock solution.

MES buffer (pH 5) was prepared as a 200 mM stock by dissolving 39 g of 2-(N-morpholino)ethanesulfonic acid (MES) in one litre dH_2O and adjusting the pH with NaOH.

2.2.2 UV/Visible Spectroscopy

Theory

UV/visible spectroscopy is commonly performed on molecules in solution, and can be used to determine electronic transitions in those molecules, [2] or to assess scattering intensity from a suspension of bacteria [3]. Irradiation with a certain wavelength (λ) of the UV/visible spectrum can lead to strong absorption of the light's energy and subsequent excitation of the molecule's valence electrons to higher energy states. The molecule is said to have an absorption maximum at this wavelength. If the absorption maximum is unknown, samples can be scanned with a range of wavelengths. Plotting the measured absorbance against wavelength reveals the characteristic absorbance peaks for the molecule(s) present in the solution. The level of absorption at the absorption maximum is proportional to the molecule's concentration, and can be used for quantification [4].

Absorption measurements are also employed to determine the turbidity in a bacterial suspension, which gives an indication of cell density and therefore bacterial number. These measurements are actually based on the amount of light scattered from the bacteria, rather than absorption of light. To maximise light scattering, UV/visible spectroscopy is performed using wavelengths that are not absorbed by the molecules in the bacterial cell.

Absorption is measured using a spectrophotometer. In general, this machine consists of two devices: a spectrometer that can produce a desired range of wavelengths of light, and a photometer that detects the amount of photons transmitted through the sample. The amount of transmitted light depends on the path length through the sample and the concentration of molecules in solution. Thus, the path length needs to be kept constant if the concentration is to be determined. The transmittance, T, is the fraction of light that passes through the sample and can be calculated as follows:

$$T = \frac{I_t}{I_0} \tag{2.1}$$

where I_t is the intensity of the transmitted light and I_0 is the intensity of the light before it passed through the cuvette. Transmittance is related to absorption, A, by the expression:

$$A = -\log(T) \tag{2.2}$$

The more light is transmitted, the lower the absorbance and vice versa. For example, if 80% of the light is transmitted through the sample, the absorption value will be approximately 0.1, whereas if only 50% of the light is transmitted, the absorption will be 0.3 (Fig. 2.2a). In the case of measuring the turbidity of a sample to give an indication of cell number, light scattering results in fewer photons reaching the detector, i.e. a loss in transmittance and therefore an increased absorbance value (Fig. 2.2b).

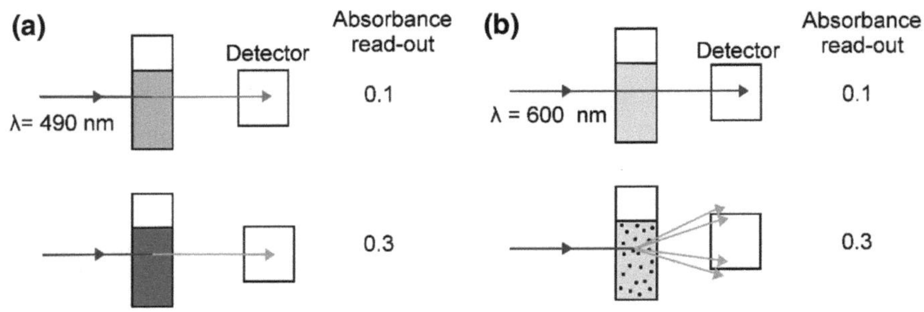

Fig. 2.2 UV/Visible spectroscopy setups. **a** Measuring absorbance using a wavelength at which the molecule has an absorption maximum. The level of absorption is proportional to the molecule's concentration. **b** Measuring the absorption in a bacterial suspension with a wavelength that is not absorbed by any components of the bacterial cell. The degree of scattering is reflected in the absorbance read-out

Method

UV/visible spectroscopy of bacterial cultures was performed in an Ultrospec 2100 Pro spectrophotometer (Amersham Biosciences, UK) using $\lambda = 600$ nm. For the measurements, 0.7 mL of background solution (sterile dH_2O or culture medium) or bacterial solution were pipetted into disposable semi-micro cuvettes (VWR, UK) and placed in the spectrophotometer. Absorbance of the background solution was subtracted from the absorbance of the bacterial culture.

All other UV/visible spectroscopy measurements were performed in microtitre plates to allow high-throughput measurement of large numbers of samples. The samples were placed in individual wells and irradiated sequentially in a SpectraMax M5e spectrophotometer (Molecular Devices, UK) using an appropriate wavelength. Background solutions were included in the measurement and subtracted from the samples' absorbance values.

2.2.3 Fluorescence Spectroscopy

Theory

Fluorescence spectroscopy is used to measure light emission from fluorophores in solution. Orbital electrons in the fluorophore are excited into high energy electronic states by the absorption of incident photons. When the electrons relax back to the ground state they lose energy through emission of a photon of longer wavelength (lower energy) than the absorbed one (Fig. 2.3a). This so-called Stokes shift is due to non-radiative relaxation, meaning that the electrons lose some of their energy through molecular vibrations, before they return to the ground state [5]. In fluorescence spectroscopy, the sample is irradiated with light of a specific wavelength (excitation). Emitted photons are collected with a detector placed perpendicular to the excitation beam (Fig. 2.3b).

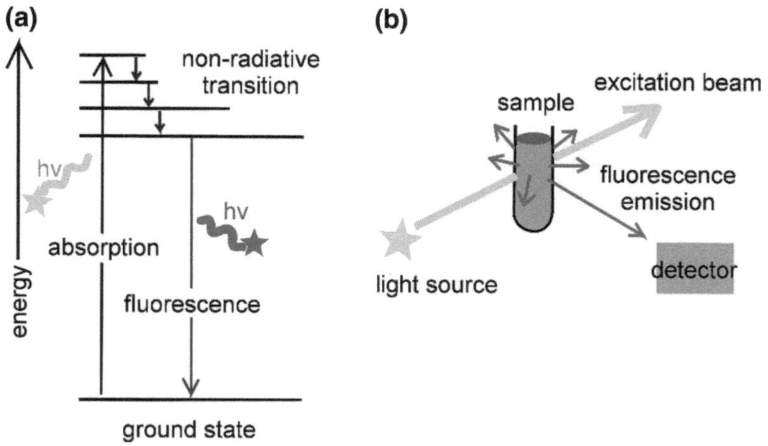

Fig. 2.3 Fluorescence spectroscopy. **a** Photons of a certain wavelength can excite the orbital electrons from a fluorophore to higher energy states. The excited electrons can lose some of their energy due to molecular vibrations, such that when they relax back to the ground state, they emit photons with a longer wavelength (less energy) than the absorbed photons. **b** Basic setup to measure fluorescence. The sample is irradiated with light of a specific wavelength (excitation), and emitted photons are collected with a detector placed perpendicular to the excitation beam

Method
Fluorescence spectroscopy measurements were performed in microtitre plates to allow high-throughput measurement of large numbers of samples. The samples were placed in individual wells and irradiated sequentially in a SpectraMax M5e spectrophotometer (Molecular Devices, UK) using an adequate excitation and emission wavelength.

2.2.4 *Bradford Assay*

Theory
The Bradford assay is a colorimetric method to determine protein concentration using the dye Coomassie Brilliant Blue G250 [6]. The dye exists in three forms, which have distinct colours: cationic (red), neutral (green) and anionic (blue). Under acidic conditions, the dye is predominantly in the doubly protonated (cationic) form, with an absorbance maximum at λ = 470 nm. When the dye is added to a protein solution it binds to basic amino acids (mainly arginine residues) and is converted to the de-protonated (anionic) form with an absorbance maximum at λ = 595 nm (Fig. 2.4). By measuring the absorbance at λ = 595 nm of a series of protein standards, the protein concentration of an unknown sample can be determined.

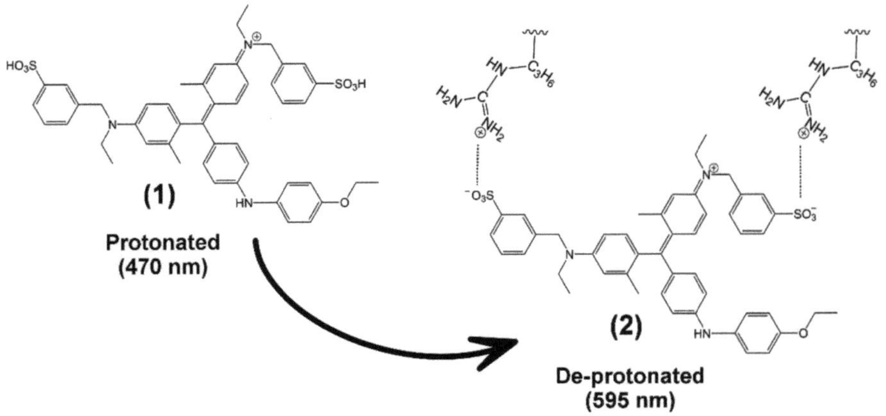

Fig. 2.4 Chemistry underlying the Bradford assay. The protonated form of the dye Coomassie Brilliant Blue G250 (*1*) predominates under acidic conditions. Binding of arginine residues on proteins causes a shift in absorbance maximum from 470 nm to 595 nm (*2*)

Method

Ferritin standards were prepared from horse spleen ferritin solution, for which the protein concentration had been determined by the supplier. Standards of 0.06, 0.125, 0.25, 0.5 and 1 mg mL^{-1} were prepared in phosphate buffer (50 mM, pH 7). Unknown samples of MF or cat-MF were diluted in phosphate buffer (50 mM, pH 7) until their colour approximately matched the colour of the 0.5 mg mL^{-1} standard. This was done to ensure the sample absorption would lie within the absorption values of the standard curve. 10 μL of each standard and unknown sample was pipetted out in triplicate into wells of a 96 well plate, 200 μL of Bradford reagent was added to each well and the plate was incubated at room temperature for 8 min. Absorbance at λ = 595 nm was measured using UV/visible spectroscopy and used to calculate the protein concentration of the unknown MF or cat-MF samples. The mean absorbance of the ferritin standards was plotted as a function of protein concentration and the slope and intercept of the linear fit were used to calculate the concentration of the unknown protein sample.

To evaluate whether cationised residues on the cat-MF surface interfered with the Bradford assay, 0.06, 0.125, 0.25, 0.5 and 1 mg mL^{-1} of MF and cat-MF were prepared in phosphate buffer (50 mM, pH 7), and the Bradford assay was carried out as described above. Absorbance values and slopes of the linear fit were similar and variations of absorbance values between MF and cat-MF were within error (Fig. 2.5). Therefore, it was concluded that cationisation did not affect the accuracy of the Bradford assay.

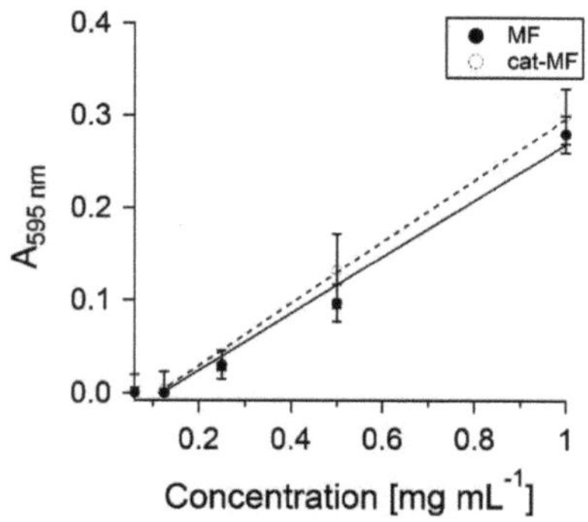

Fig. 2.5 Bradford assay carried out on MF and cat-MF samples (n = 2). The slopes of the linear fit are 0.31 for MF (*solid line*), and 0.33 for cat-MF (*dashed line*). Absorbance values are similar for MF and cat-MF

2.3 Chromatography Methods

Chromatography is a technique for the separation of a mixture of components in a solution or suspension. The mixture is passed through a medium in which the components interact more or less strongly with the medium depending on particular properties, such as charge or size. Therefore, they move through the medium at different rates and can be collected in sequential fractions. Here, two chromatography techniques were used to purify MF after synthesis. During MF synthesis, nanoparticles form inside the ApoF cavity, as well as in the buffer solution. Thus, ion exchange chromatography was employed to separate MF from the free nanoparticles. Agglomerates of MF were then separated from MF monomers using size exclusion chromatography. All chromatography methods were performed using an ÄKTApurifier system (GE Healthcare, UK).

2.3.1 Ion Exchange Chromatography

Theory

Ion exchange chromatography (IEC) separates proteins based on their charge [7]. Reversible interactions between charged molecules and the oppositely charged ion exchange matrix are controlled to achieve binding or elution of particular molecules, resulting in their separation. The ion exchange matrix consists of porous, spherical particles substituted with ionic groups that are negatively or positively charged. The pH plays an important role in IEC of proteins, because proteins have no net charge at a pH equivalent to their isoelectric point and therefore won't interact with a charged medium. However, at a pH above their isoelectric point, proteins will bind to a positively charged matrix and at a pH below it proteins bind to a negatively charged matrix. Both pH and ionic strength of the buffer are selected such that proteins of interest bind to the matrix when the sample is loaded, while unwanted material does not bind and can be washed out with buffer (Fig. 2.6a, b). To elute the bound protein the buffer conditions are changed, for example by

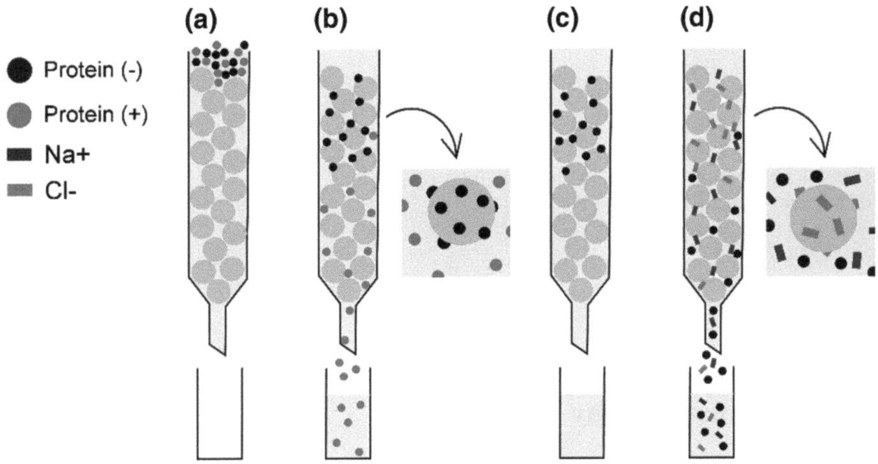

Fig. 2.6 Schematic of ion exchange chromatography. **a** A mixture of proteins with different charges at a given pH is loaded onto the column containing a charged matrix. **b** Proteins with a charge opposite to that of the matrix are immobilised, while all other components are eluted. **c** A wash step is included to remove any material that is not immobilised, such that only the target protein remains trapped in the column. **d** Elution of the target protein is achieved by increasing the ionic strength of the buffer. As ionic strength increases, the salt ions compete with the bound proteins for charges on the matrix surface. Bound molecules are detached, move down the column and can be collected

increasing the ionic strength (salt concentration) of the buffer. As ionic strength increases, the salt ions (typically Na^+ or Cl^-) compete with the bound proteins for charges on the matrix surface such that bound molecules detach and move down the column (Fig. 2.6d). The components with the lowest net charge at the selected pH will elute first from the column, while higher charged species will be more strongly bound and elute as ionic strength increases. Results of IEC are displayed as an elution profile generated by a UV/visible spectroscopy cell incorporated into the purifier setup. The absorbance of the solution is monitored at a particular wavelength (e.g. 280 nm) as it exits the column. Peaks in the absorbance indicate when proteins are eluted.

Method

Samples were loaded onto a column (2.5 cm diameter, 20 cm long) containing the cationic matrix ANX Sepharose 4 Fast Flow (GE Healthcare, Life Sciences) using a P960 sample pump (GE Healthcare, UK) at a flow rate of 10 mL min^{-1}. Tris buffer (50 mM, pH 8) was used as a running buffer (flow rate 10 mL min^{-1}), with which metal oxide nanoparticles were eluted from the column, while protein nanoparticles were immobilised on the matrix. To elute the protein, Tris buffer containing increasing concentration of NaCl was added (from 100 to 1000 mM) at the same flow rate. The protein eluted at a NaCl concentration of 500 mM and was collected in 50 mL fractions using an automated fraction collector (Frac-950, GE Healthcare, Life Sciences).

2.3.2 *Size Exclusion Chromatography (SEC)*

Theory

Size exclusion chromatography (SEC) separates the components in a sample based on differences in size as they pass through a filtration medium packed inside a column [8]. The filtration medium is a porous matrix that consists of spherical particles (Fig. 2.7a). The matrix is equilibrated with buffer, which fills all pores and spaces. Thus, it forms a stationary phase that is in equilibrium with the buffer that is pumped through the column (mobile phase). Unlike IEC, molecules passed through the column in the mobile phase do not bind to the filtration medium, but are eluted from the column in order of their molecular size. Larger molecules that cannot enter the pores of the matrix are eluted first, while smaller molecules with partial access to the pores elute in order of decreasing size (Fig. 2.7b, c). Very small molecules, such as salts, have full access to the pores and move down the column very slowly. Results from SEC are displayed as an elution profile (see Sect. 2.3.1).

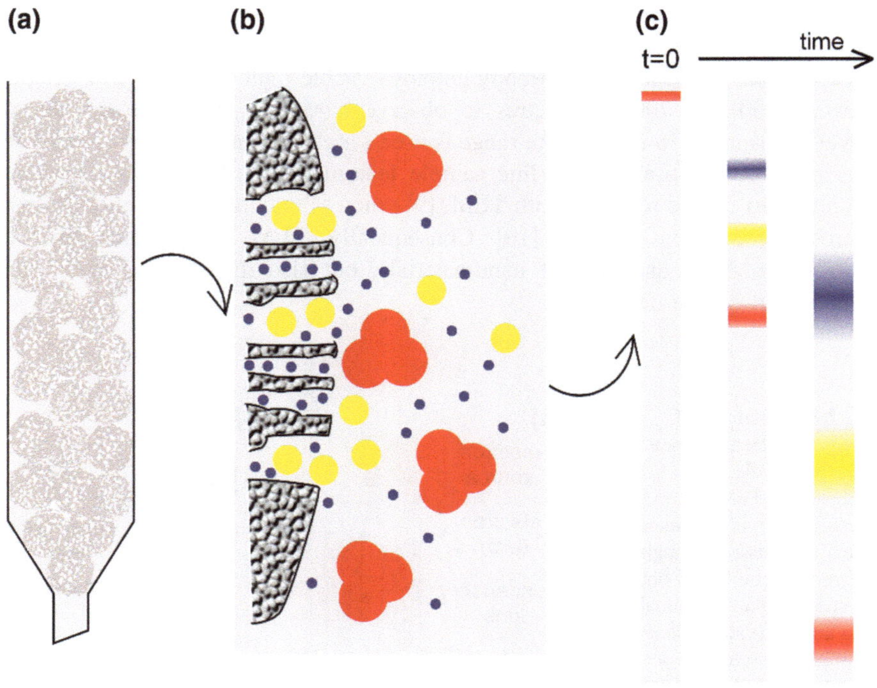

Fig. 2.7 Size exclusion chromatography. **a** A column is packed with a matrix of spherical, highly porous particles. **b** When the sample is passed through the column matrix, components in the sample can access pores in the matrix according to their size. **c** Schematic of the gel filtration process over time. The sample is loaded onto the column at t = 0. Over time, the components of the sample pass through the length of the column and are separated according to their sizes (*blue* smallest molecules, *red* largest molecules, *yellow* middle-sized molecules)

Method

MF samples pre-purified using IEC were concentrated using Centriprep centrifugal filter units (Merck Millipore, UK) to a volume of approximately 2 mL. They were loaded onto a HiPrep 26/60 Sephacryl S-300 h column (GE Healthcare, Life Sciences) column using an injection loop. Tris buffer (50 mM, pH 8.0) containing 150 mM NaCl was used as running and elution buffer at a flow rate of 1.6 mL min^{-1}. Fractions of 6 mL volume were collected using an automated fraction collector (Frac-950, GE Healthcare, Life Sciences).

2.4 Characterisation Methods

2.4.1 Transmission Electron Microscopy

Theory

The working principle of TEM is similar to that of light microscopy in that the image of a specimen is magnified and focussed onto an observation plane (Fig. 2.8a, b). The crucial difference is the type of energy beam used to "illuminate" the specimen: while light microscopy employs visible light beams of wavelengths between 400 and 700 nanometres to observe a sample, an electron beam of wavelengths in the low picometre range is used for electron microscopy. Therefore, it is possible to observe very fine sample features using TEM. Features about 0.2 nm apart can be resolved with TEM, [9] whereas maximum resolution in light microscopy is about 200 nm [10]. Consequently, TEM is frequently used to characterise shape and size of nanomaterials, but also their crystal phase and composition can be determined.

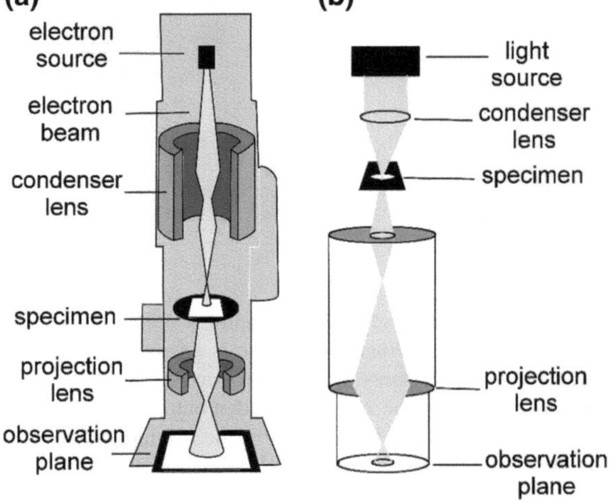

Fig. 2.8 Schematic of microscopic techniques. **a** In a transmission electron microscope (TEM) an electron beam is generated, which is focussed through electromagnetic lenses onto the specimen and eventually onto an observation plane on a fluorescent screen or a digital camera. **b** Analogous, in a light microscope a visible light beam is focussed through glass lenses onto a specimen and then onto an observation plane

The most common mode of operation in TEM is the bright field imaging mode. In this mode, the contrast of the TEM image is due to absorption and scattering of electrons in the sample. Thick samples or materials with a high atomic number scatter electrons more strongly and will appear dark, while regions without sample in the beam path appear bright (hence the name "bright field"). However, the elements that make up organic materials, such as proteins and lipids, have very low atomic numbers and are therefore "invisible" in a TEM image. To visualise these types of materials, a variety of metal-based stains are employed. These can be negative stains, which surround organic materials such that they appear as bright structures on a dark background. For example, the protein capsules of viruses are commonly negative stained with neutralised phosphotungstic acid [11]. Stains can also bind to specific biological structures like phospholipid membranes, proteins and nucleic acids. To visualise these structures in cell sections, the cells should first be fixed to avoid structural changes occurring during the staining procedure. Common fixatives are glutaraldehyde, which cross-links proteins in the cell through methylene bridges, [12] as well as osmium tetroxide, which mainly reacts with the carbon-carbon double bond in unsaturated fatty acids chains in cell membranes [13]. Uranyl acetate is routinely used for contrast staining of biological structures because uranyl ions bind to carboxyl groups of proteins and lipids, and to phosphate groups of nucleic acids. Therefore, uranyl acetate staining achieves good contrast staining of membranes, nucleic acids and protein complexes associated with nuclei acids, such as ribosomes [14]. Staining with lead citrate enhances the contrast for several cellular structures such as phospholipid membranes, ribosomes and the cytoskleleton, particularly after osmium tetroxide fixation [15].

Methods

A variety of preparation methods was used in this work depending on the sample to be imaged.

To confirm **nanoparticle mineralisation** and determine MF core size and overall diameter, 10 μL samples containing approximately 0.5 mg mL^{-1} protein were placed onto carbon-coated copper grids (Agar Scientific, UK) for 10 min. The excess was removed with filter paper and either left to dry or stained with 2% (w/v) neutralised phosphotungstic acid (pH 7) for 30 s before air drying. Samples were observed using an acceleration voltage of 120 kV on a JEM 1200 EX transmission microscope (JEOL, UK) and images recorded with a MegaView II digital camera (Olympus, UK) running analySIS 3.0 software (Soft Imaging Systems GmbH, Germany).

To observe **stem cells** exposed to cat-MF, cells were fixed with 2.5% (w/v) glutaraldehyde in 0.1 M cacodylate buffer (pH 7.4) and with 0.5% (w/v) osmium tetroxide in cacodylate buffer, each for 30 min at room temperature. Fixed cells were washed in cacodylate buffer and distilled water, before dehydration in a graded series of ethanol (from 70% to 100%). After that, they were embedded in an epoxy resin (Epon), and the resin polymerised at 60 °C for 24 h. Thin sections of 70 nm thickness were cut from the resin blocks and transferred onto carbon-coated copper grids. Sections were contrast stained with 3% (w/v) uranyl acetate and 0.5% (w/v) alkaline lead citrate. Localisation of cat-MF on labelled cells was investigated using a Tecnai 12 BioTwin Spirit transmission electron microscope (FEI, UK) with an acceleration voltage of 120 kV. Images were recorded with an Eagle 4 k camera (FEI) running TEM Imaging and Analysis software (FEI).

To image **bacteria** exposed to cat-MF, 10 μL samples of bacterial suspension were placed onto carbon-coated copper grids for 10 min. The excess was removed with filter paper and grids were floated on 100 μL drops of 5% (v/v) glutaraldehyde solution for 30 min at room temperature to fix the cells. After that, grids were washed by placing them on three 100 μL drops of dH_2O for 5 min each and air dried. Samples were observed using an acceleration voltage of 120 kV on a JEM 1200 EX transmission microscope and images recorded with a MegaView II digital camera running analySIS 3.0 software.

2.4.2 *Dynamic Light Scattering*

Theory

DLS is frequently used to assess the hydrodynamic diameter of micro- and nanometre-sized particles in aqueous suspensions [16]. In this technique, the particle size is inferred from the extent of Brownian motion the particles display: small particles experience a higher degree of Brownian motion compared to larger ones. The motion of the particles is assessed using a laser beam which is directed through the sample. The light from the beam is scattered by the particles, such that a scattering pattern of fluctuating intensity can be observed using a detector at a particular angle to the incoming beam (Fig. 2.9a). The temporal change of this intensity fluctuation is observed, and a correlation function calculated. For large, slow moving particles, this pattern will change more slowly, and thus the rate of decay of the correlation function will be smaller, compared to smaller, fast moving

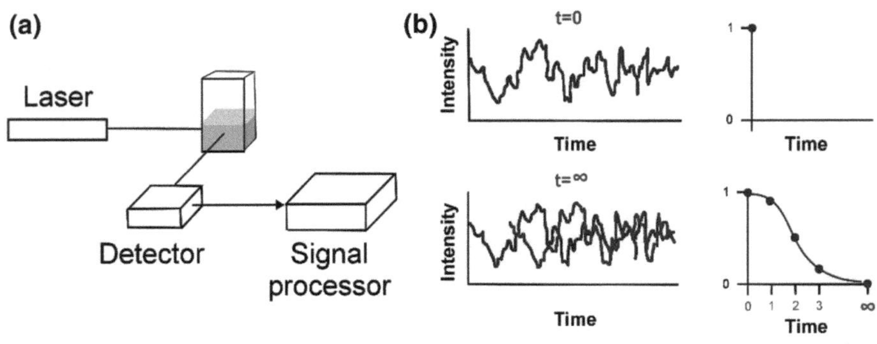

Fig. 2.9 Schematic of a dynamic light scattering experiment. **a** A sample is illuminated with laser light and the intensity pattern that is created by scattering from particles in the sample is observed with a detector. **b** Due to particle movement, the intensity pattern changes over time. Correlation of the pattern at a given time (t = 1, 2, 3, …) with the pattern at the start of the measurement (t = 0) gives rise to a correlation function, from which the hydrodynamic radius can be calculated

particles (Fig. 2.9b). Thus, the correlation function gives an indication of the particle's speed due to Brownian motion, which in turn relates to the particle size [17]. This relationship is described by the Stokes-Einstein equation:

$$d = \frac{k_B \cdot T}{3 \cdot \pi \cdot \eta \cdot D} \quad (2.3)$$

where d is the hydrodynamic diameter of the particles, k_B is the Boltzmann constant, T is the absolute temperature, η is the viscosity of the medium in which the particles are suspended, and D is the diffusion coefficient, which can be calculated from the correlation function. By collecting several measurements for any one sample, a mean particle size, the **z-average**, and an estimate of the width of the size distribution, the **polydispersity index**, can be calculated.

Although this technique is very simple and powerful, care must be taken in the interpretation of the results, particularly if comparisons between samples are to be made. Because the result is determined through assessing particle motion, factors that influence this motion (e.g. temperature, ionic strength of the solution) can also influence the result and therefore need to be kept constant. It also needs to be considered that larger particles scatter light more strongly than small articles, such that a small amount of larger particles in a mixture of sizes can dominate the signal and therefore the calculated z-average. It is advisable to reduce contamination of the sample with large particles, such as dust and aggregates, by filtering or centrifuging the sample prior to DLS measurements.

Method

Dynamic light scattering (DLS) measurements were performed on a ZetaSizer Nano-ZS (Malvern Instruments, UK). All samples were suspended in 20 mM phosphate buffer at room temperature and pH 7, and adjusted to a protein concentration of approximately 1 mg mL^{-1} before passing them through a filter of 220 nm pore size. 1 mL of each solution was placed in folded capillary tubes (Malvern Instuments, UK). A total of 13 to 15 measurements was collected for each sample.

2.4.3 Zeta Potentiometry

Theory

Charged particles in solution are surrounded by an electrical double layer of ions present in the solvent. In a dynamic particle system, ions are only retained within a theoretical plane of hydrodynamic shear, and the potential energy at this plane is referred to as the zeta potential (Fig. 2.10a) [18]. Particles with predominantly positive charges in their surface have a positive zeta potential, and negatively charged particles a negative one. The zeta potential is determined by measuring the electrophoretic mobility of a charged particle in an applied electric field (Fig. 2.10b). A laser is used to irradiate a sample and the scattered light is collected

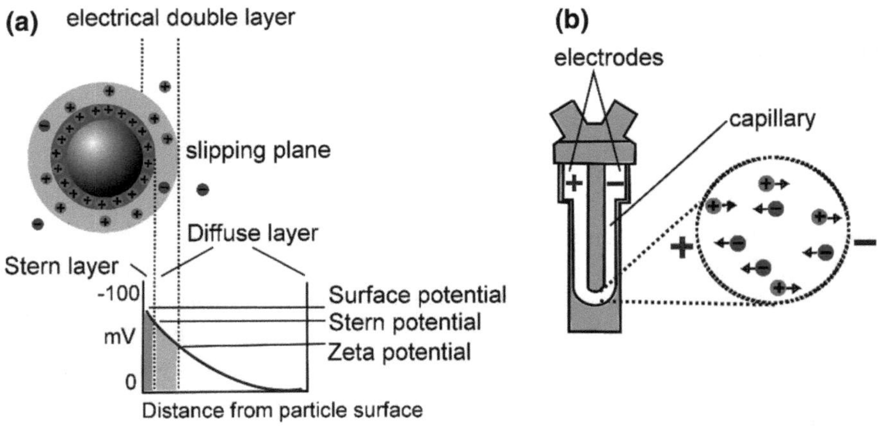

Fig. 2.10 The zeta potential of a charged particle. **a** Electrical double layer surrounding a charged particle in an electrolyte solution. **b** The folded capillary tube used to measure the electrophoretic mobility of charged particles in an applied electric field, from which the zeta potential can be calculated

as a function of time. The electrophoretic mobility is determined by comparing the phase of the scattered light with the phase of a reference beam [19]. The Henry equation relates electrophoretic mobility, *EM*, of a particle in a liquid medium to its zeta potential, *ZP*:

$$EM = \frac{2 \cdot \varepsilon \cdot ZP \cdot f(ka)}{3 \cdot \eta} \quad (2.4)$$

where ε is the dielectric constant of the medium, *f*(*ka*) is the Henry's function, and η is the viscosity of the medium. The Henry's function describes the ratio of the particle's radius to the thickness of the electric double layer surrounding it. In general, either 1.5 or 1.0 are used as approximations of *f*(*ka*). For measurements of particles suspended in aqueous media with moderate electrolyte concentration (as is the case in this work), 1.5 is used for *f*(*ka*), and is referred to as the Smoluchowski approximation.

The most important factor that affects zeta potential is the pH of the solution, because it influences surface charge through protonation/deprotonation of functional groups. Therefore, zeta potential values must be quoted with the pH at which they were measured, otherwise they become an almost meaningless number.

Method

Zeta potential measurements were performed on a ZetaSizer Nano-ZS (Malvern Instruments, UK). Samples were suspended in 20 mM phosphate buffer at pH 7, and the protein concentration was adjusted to approximately 1 mg mL^{-1} and passed through a 220 nm filter. 1 mL of sample solution was placed in folded capillary tubes, and 10 to 12 measurements were performed on each sample.

2.4.4 Mass Spectrometry

Theory

Mass spectrometry is a technique to determine the molecular mass of large biomolecules, as well as small organic molecules. The measurements rely on the ionisation of the sample molecules, followed by their extraction into an analyser region, in which they are separated according to their mass-to-charge ratio (m/z). A detector monitors the ion current from the separated ions, and the intensity of the ion current can be plotted against m/z values to yield an m/z spectrum. This way, the number of components in the sample, and the molecular weight and relative abundance of each component can be determined.

In this work, matrix-assisted laser desorption ionisation time-of-flight (MALDI-TOF) mass spectrometry was used. This is a common technique to analyse samples of high molecular mass, such as proteins [20]. The sample is mixed with an excess of matrix compound (e.g. 2,5- dihydroxybenzoic acid) in a volatile solvent (e.g. methanol) that evaporates to yield a solid, crystalline matrix. This matrix strongly absorbs incident laser light, which results in thermal ablation. The sublimating matrix carries ionised sample molecules into the gas phase, where they are all accelerated to the same kinetic energy using an applied electric field. Accelerated ions are injected into a time-of-flight analyser, in which they are separated according to their m/z by measuring the time it takes for ions to move through a flight tube (heavier ions moving slower than lighter ones) (Fig. 2.11) [21]. The m/z scale can be calibrated using a sample of known molecular mass. MALDI spectra are relatively easy to interpret, because the gentle ionisation

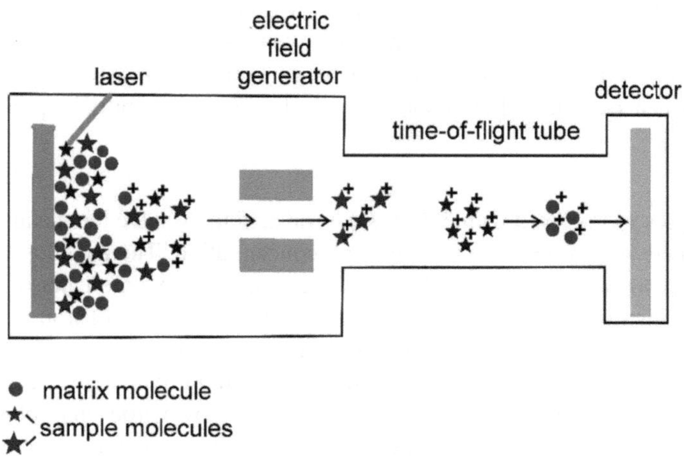

Fig. 2.11 Schematic of a MALDI-TOF experiment. The sample is embedded in a matrix, which strongly absorbs incident laser light. This causes thermal ablation of matrix and sample molecules, resulting in their ionisation and sublimation into the gas phase. An applied electric field accelerates the ions to the same kinetic energy. Accelerated ions are injected into a time-of-flight tube where they are separated according to their mass

method results in the generation of predominantly singly charged ions regardless of the molecular mass. This avoids multiple traces of the same molecule due to multiply charged sample ions.

Method

The MALDI-TOF experiment was performed on ApoF and cat-ApoF samples, instead of mineralised MF and cat-MF samples, because the mineralised protein has an exceedingly high molecular mass (over 900 kDa) due to the metal core. ApoF and cat-ApoF were dissolved in an equivalent volume of 20 mg mL^{-1} of 2,5-dihydroxybenzoic acid in methanol. Samples of 2 µL were pipetted onto a steel plate and left to air dry and analysed in a Bruker UltrafleXtreme (Applied Biosystems, UK) running FLEX Control software (Bruker, USA).

2.4.5 Inductively-Coupled Plasma Optical Emission Spectroscopy

Theory

Inductively-coupled plasma optical emission spectroscopy (ICP-OES) is frequently used to quantify the elements present in a variety of samples, such as nanoparticles, drinking water, agricultural and biological samples [22]. This methods relies on the fact that vaporised atoms can be excited by an inductively-coupled argon plasma and emit photons of a characteristic pattern of wavelengths as they return to the ground state. The wavelengths in the emitted spectrum are determined by the energy level structure of the atoms, thus, they can be used to identify the elements from which they originated. The amount of emitted light is proportional to the amount of atoms present in the sample, and by introducing known concentrations of analyte atoms, a calibration curve can be generated. Using ICP-OES, up to 70 elements can be detected simultaneously at low parts per billion (ppb) detection limits [22].

Method

MF samples were mixed with 50% (v/v) nitric acid to dissolve the metal atoms, and then diluted with water to reduce the acid concentration. The aqueous sample was introduced as an aerosol into the central channel of an argon plasma in a 710 ICP-OES instrument (Agilent, UK). To quantify the amounts of iron and cobalt in the MF samples, optical emission was measured at specific wavelengths characteristic for the emitted spectrum of these elements (approximately 255 nm for iron and 233 nm cobalt). Five calibration samples of known iron and cobalt concentrations were analysed alongside the MF samples. The protein concentration in the MF samples was determined using the Bradford assay (see Sect. 2.2.4), such that iron content per protein unit could be calculated.

2.4.6 SQUID Magnetometry

Theory

SQUID magnetometry uses a superconducting quantum interference device (SQUID) for the detection of magnetic moments in liquid or solid samples. It is a very sensitive technique, which allows detection of magnetic moments as small as 10^{-11} Am^2 [23]. The magnetic moment can be measured as a function of an applied field or an applied temperature. Hence, it is a common technique used to characterise important parameters of magnetic nanoparticles, such as saturation magnetisation and magnetic susceptibility. Both can be determined by measuring the magnetic moment as a function of an applied magnetic field. The blocking temperature can be estimated by measuring the magnetic moment as a function of the applied temperature. A measurement is performed by moving a sample through superconducting pick-up coils, by which a current is induced in the coils (Fig. 2.12a). This is converted into a voltage signal by the SQUID, with the amplitude of the generated voltage curve being proportional to the magnetic moment of the sample (Fig. 2.12b) [23, 24].

Method

Saturation magnetisation and magnetic susceptibility of MF and cat-MF were measured by performing a field sweep over ± 2 T at 300 K using a SQUID magnetometer (Magnetic Property Measurement System, MPMS, Quantum Design, USA). A 0.2 mL tube (Corning, UK) and a drinking straw were used as a sample holder. The contribution of the sample holder to the overall sample moment was

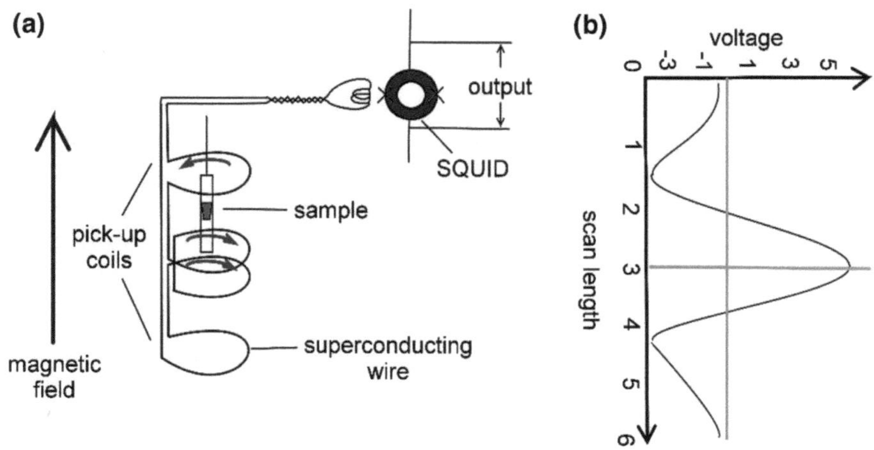

Fig. 2.12 Detection of small magnetic moments using a SQUID magnetometer. **a** A sample's magnetic moment in an applied magnetic field is measured by moving it through pick-up coils made from superconducting wire. The induced current is converted to a voltage signal using a superconducting quantum interference device (SQUID). **b** The voltage output as a function of sample position in the pick-up coils. The magnetic moment can be calculated from this voltage curve

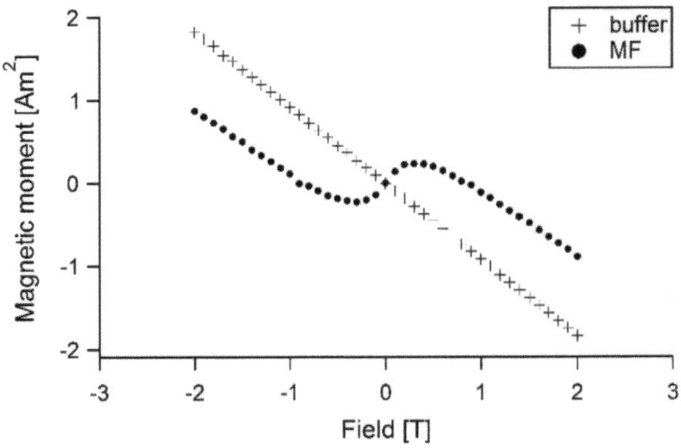

Fig. 2.13 Assessing the diamagnetic background. The diamagnetic contribution to the overall MF sample moment can be mainly attributed to the phosphate buffer

measured separately and the background-corrected sample moment was determined as described in Appendix B. Briefly, a corrected voltage curve for MF/cat-MF was computed by subtracting the sample holder curve from the curve recorded for the sample and the sample holder together using a MATLAB (MathWorks, USA) script written by Dr. Carsten Putzke (University of Bristol). The best fit for this curve and the magnetic moment were found using equations supplied by the manufacturer.

The diamagnetic background from the aqueous solvent (50 mM phosphate buffer) was measured separately and found to be a major contributor to the diamagnetic response of the MF/cat-MF sample, particularly in high fields (Fig. 2.13). To correct for this diamagnetic contribution, a linear slope was subtracted from the magnetic moment of the sample.

Superparamagnetic behaviour was confirmed by fitting a Langevin function to the measured data points:

$$m\,(B) = \mu * n * L(x) \tag{2.5}$$

where $m(B)$ is the magnetic moment of the sample measured as a function of the applied field B, μ is the magnetic moment of an individual nanoparticle, n is the number of nanoparticles in the sample, and $L(x)$ is the Langevin function:

$$L\,(x) = \coth\,(x) - x^{-1} \tag{2.6}$$

with

$$x = \frac{\mu B}{k_B T} \tag{2.7}$$

where k_B is the Boltzmann constant and T is the absolute temperature. The saturation magnetisation is given by μ*n. Magnetic susceptibility was calculated by determining the slope of the linear part of the magnetisation curve in the low field region (up to ± 0.3 T).

The blocking temperature was determined by cooling the sample from 120 K to 6 K in zero field, then applying a field of 0.01 T and measuring the magnetic moment of the sample at intervals of 1 K up to 25 K, and of 2 K up to 120 K.

2.4.7 Magnetic Resonance Imaging

Theory

When placed inside the magnetic coil of a magnetic resonance imaging (MRI) scanner, the magnetic moments of hydrogen nuclei (protons) in aqueous solutions align with the external magnetic field generated by the coil. The magnetic moments of these protons arise from their nuclear spins, which precess around the axis of the external field at a characteristic frequency, the so called Larmor frequency. The sum of all these spins results in a net magnetisation (M_O) that is aligned with the external field but of a much smaller magnitude (of the order of μT) (Fig. 2.14a). While the magnetisation is at equilibrium (lying parallel to the external field), it is not possible to measure it directly. However, radiofrequency (RF) pulses that are applied at the Larmor frequency (the protons' resonance frequency) can impart energy on the protons that flip the orientation of their spins 90° out of their longitudinal alignment with the external field into the transverse plane. Detector coils that are sensitive to only the transverse magnetisation can then pick up this magnetic resonance (MR) signal [25]. Thus, M_0 becomes a significant signal that can be recorded. This is done by creating echoes, which will be discussed below. When the RF pulse is removed, the protons can relax back to their initial state of equilibrium. There are two relaxation mechanisms, spin-spin and spin-lattice relaxation [26]. During spin-spin relaxation, neighbouring protons interact such that their precession frequencies gradually de-phase, and thus the MR signal measured in the transverse plane decays exponentially (Fig. 2.14b, top). Therefore, spin-spin relaxation is also called transverse relaxation. During spin-lattice relaxation, the protons transfer the energy they received from the RF pulse to the surrounding environment (the lattice) and thus gradually return to their equilibrium state, recovering their initial longitudinal magnetisation (Fig. 2.14b, bottom). Therefore, spin-lattice relaxation is also called longitudinal relaxation. When imaging humans or animals with MRI, the transverse and longitudinal relaxation times depend on the composition of the tissues, and each tissue type has a set of characteristic time constants associated with those processes. T1 describes the time it takes for the spins to recover 63% of their longitudinal magnetisation, while T2 describes the time it takes for the transverse magnetisation to drop to 37% of its initial value.

Both T1 and T2 can be influenced by introducing contrast agents, which can aid MRI based diagnoses [27]. The effect of contrast agents on relaxation times of individual tissues is described using relaxation rates, which are the inverse of the relaxation times, $R_1 = T1^{-1}$ and $R_2 = T2^{-1}$. For contrast agents, a specific relaxivity rx (x = 1, 2) is defined, which describes how much they change T1 and T2 per molar concentration.

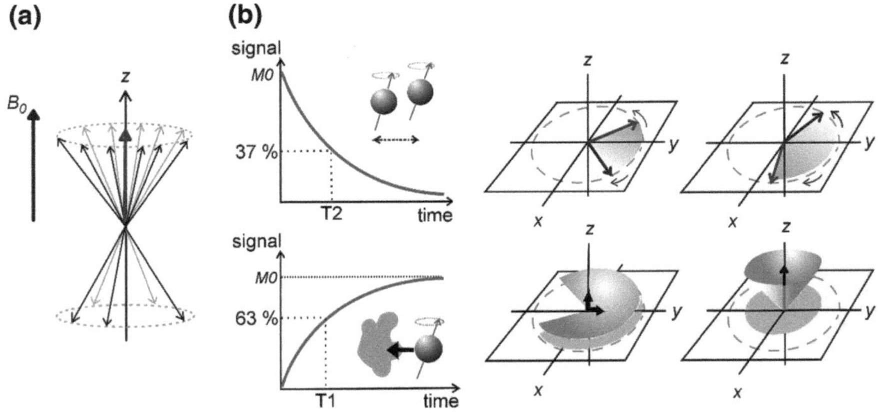

Fig. 2.14 MRI relaxation mechanisms. The nuclear spins of protons in an applied field which precess around the axis of the external field (B_O) at the Larmor frequency. The sum of all spins results in a net magnetisation (M_O) that is aligned with B_O (*blue arrow*). A radiofrequency (RF) pulse that is applied at the Larmor frequency can impart energy to the protons and flip the orientation of their spins 90° out of their longitudinal alignment with the external field into the transverse plane. **b** Relaxation mechanisms after the RF pulse is removed. Top: during spin-spin relaxation, neighbouring protons interact such that their precession frequencies gradually de-phase (illustrated by a "fanning out" of the arrows). Thus the MR signal measured in the transverse plane decays exponentially. Bottom: during spin-lattice relaxation, the protons transfer the energy they received from the RF pulse to the surrounding environment and thus gradually recover their initial longitudinal magnetisation

As mentioned earlier, the MR signal and its temporal change can be measured by creating echoes. Two types of echoes are mainly used in MRI: gradient echo and spin echo. In a spin echo sequence, proton spins are left to de-phase for a certain amount of time after the 90° pulse. Then, a pulse is applied that flips all spins through 180° about the y axis, reversing their phase angles. This means that after the same time, all spins will re-phase to form a spin echo (Fig. 2.15). The echo height, and thus signal intensity, will mainly depend on the transverse relaxation time T2. Therefore, this is a useful sequence to measure T2 [25].

The gradient echo sequence also starts with an RF excitation pulse, usually with a flip angle smaller than the 90° used in the spin echo sequence. Instead of letting the spins de-phase naturally as in the spin echo, a negative magnetic gradient is applied immediately after the excitation pulse, which causes rapid de-phasing of the spins. After the negative gradient, a positive gradient is applied, which reverses the magnetic field gradient. This means that spins precessing at low frequencies due to their position in the gradient now precess at a higher frequency, and vice versa. Therefore, the spins start to re-phase and eventually all come back into phase along the y-axis, forming the gradient echo. However, the positive gradient only compensates for the de-phasing caused by the initially applied negative gradient. It does not re-focus the spins that have de-phased due to magnetic field inhomogeneities. Therefore, this imaging mode is very useful to image contrast enhancement caused

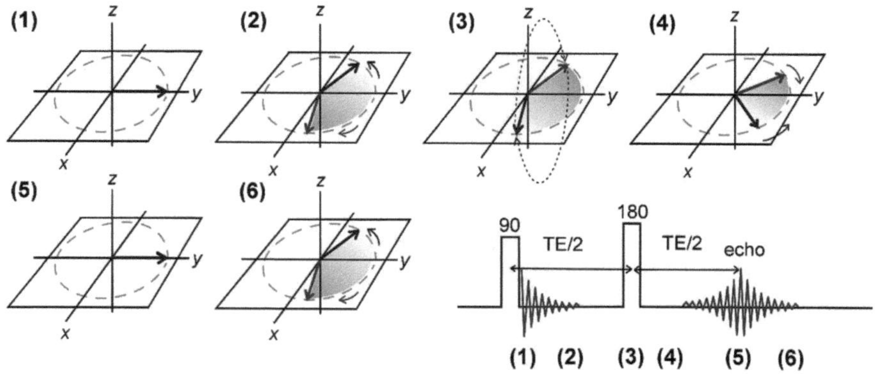

Fig. 2.15 Schematic of a spin echo sequence. An RF pulse flips the magnetisation of the protons into the transverse plane (*1*), and when the pulse is removed the spins de-phase (*2*). Another RF pulse is applied that flips all spins through 180° about the y axis, reversing their phase angles (*3*). As a result, spins start to re-phase (*4*) and eventually come into phase again (*5*) before de-phasing again (*6*)

by SPIONs, because SPIONs behave like small magnets in the tissue creating significant magnetic field inhomogeneities.

To measure T1 a series of inversion-recovery (IR) spin echo sequences with varying inversion delay times can be used [28]. IR sequences start with a 180° pulse that inverts the magnetisations of the spins, followed by a delay time (TI) during which T1 recovery occurs. When the 90° pulse is applied, which flips the spins in the transverse plane for detection, some of the signals may still be negative (inverted), resulting in a mixture of positive and negative echoes that are represented as light and dark voxels in the image. The repetition time (TR) between IR sequences should be at least five times the longest T1 to allow relaxation to the initial equilibrium magnetisation (M_0) after the 90° pulse (Fig. 2.16).

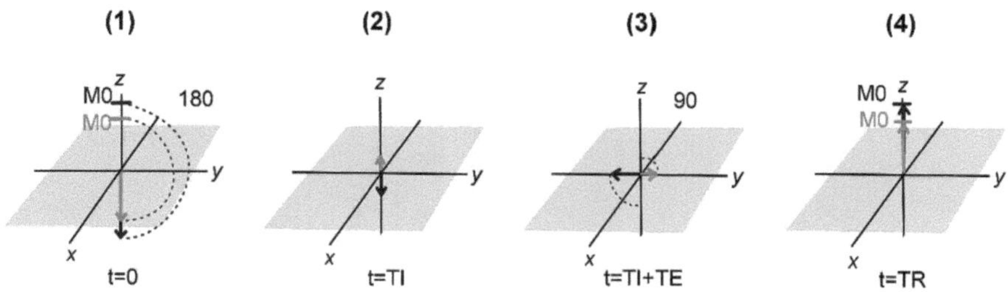

Fig. 2.16 Schematic of an inversion-recovery spin echo sequence. An initial 180° pulse inverts the magnetisations of the proton spins (*1*) (the two arrows represent spins in two different environments). During the delay time t = TI, T1 recovery occurs (*2*). When the 90° pulse is applied some of the signals may still be inverted, resulting in a mixture of positive and negative echoes (spatially distributed across the different environments) (*3*). The repetition time, TR, must always be at least five times the longest T1 to allow full relaxation (*4*)

Method

To determine specific relaxivities of MF and cat-MF, T1 and T2 were measured for protein concentrations ranging from 3 μM to 0.05 μM in phosphate buffered saline (PBS). 200 μL of each solution were placed in tubes of a PCR plate (Corning, UK) embedded in a 1.5% (w/v) carrageenan gel. T1 was determined using a series of IR spin echo sequences with varying TI. A TR of 6000 ms and an echo time of 9.6 ms was chosen. Signal intensity after the following TI was recorded: 50, 150, 250, 350, 450, 650, 850, 1250, and 2700 ms (slice thickness: 3 mm, field of view: 12 cm). The signal intensity, *S*(*TI*), at each TI is given by:

$$S(TI) = M_0\left[1 - 2\exp\left(-\frac{TI}{T1}\right)\right] \quad (2.8)$$

To measure T2, a series of spin-echo sequences was used and the signal intensities were measured after the following TE: 10, 20, 30, 50, 80, 100, 160, 240, 480 ms (slice thickness: 3 mm, field of view: 12 cm. Because transverse relaxation occurs faster than longitudinal relaxation, a shorter repetition time was chosen (TR = 3500 ms). The signal intensity, *S*(*TE*), at each TE is given by:

$$S\,(TE) = M_0 \exp\left(-\frac{TE}{T2}\right) \quad (2.9)$$

Signal intensities in the generated greyscale images were determined by region of interest (ROI) image analysis using a program written in Python (Python Software Foundation, Netherlands) by Ronald Hartley-Davies (University Hospitals Bristol NHS Foundation Trust). ROIs were determined by identifying the circular regions in the image, which represented the diameters of the PCR tubes containing the samples. Mean intensity of the pixels within the circular regions calculated by

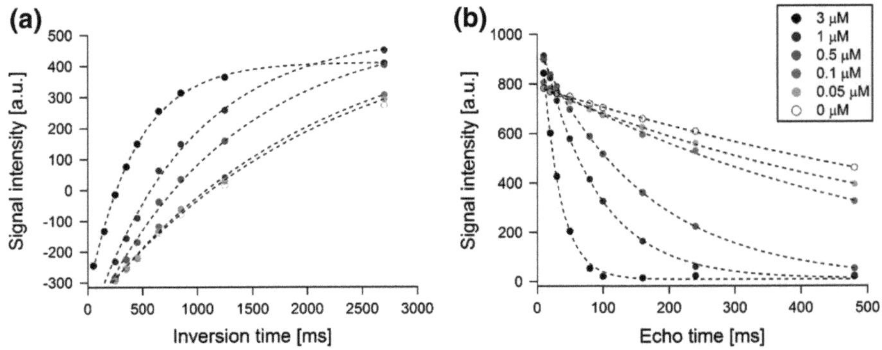

Fig. 2.17 Measuring T1 and T2 of MF. **a** A series of inversion recovery spin echo sequences with varying inversion times was performed for each MF concentration, and signal intensities were measured after each inversion time. **b** A series of spin echo sequences was performed and signal intensity was measured after varying echo times. The same data were also recorded for cat-MF

the software were plotted against the TI or TE, respectively, and T1 and T2 were determined by fitting exponential curves to the signal intensities according to Eqs. 2.8 and 2.9 (Fig. 2.17). The calculated T1 and T2 values were then plotted against the molar iron concentration present in the samples as determined by ICP-OES. The slope of the linear fit was determined to be the relaxivity, r1 and r2, respectively.

2.5 Magnetoferritin Synthesis

Cobalt-doped MF was synthesised from commercially available horse spleen ApoF using an established protocol [29, 30]. Synthesis was performed in a double-jacketed reaction vessel at 65 °C containing 75 mL of deoxygenated 50 mM HEPES buffer (pH 8.6) with 3 mg mL^{-1} ApoF. The vessel was kept under a nitrogen atmosphere to restrict oxidation of the metal precursors. 30.3 mL of 25 mM ammonium iron sulphate hexahydrate, containing 2.5% (w/v) cobalt sulphate heptahydrate was added at a rate of 0.15 mL min^{-1} with a syringe pump (Titrando 907, Metrohm). Controlled oxidation of the metal precursors inside the ApoF cavity was mediated by adding an equivalent volume of an 8.33 mM hydrogen peroxide solution at the same rate. Fresh metal precursor and hydrogen peroxide solutions were injected three times at 65 min intervals. The initially colourless ApoF solution adopted a dark brown colour upon nanoparticle mineralisation, and after the final reagent addition it was left to mature for 15 min before 1.5 mL of 1 M sodium citrate were added to chelate free metal ions in the solution. Large precipitates were removed by centrifuging the sample for 30 min at 4350 × g and then passing the supernatant through a 220 nm syringe filter. The protein was purified by anion-exchange chromatography (see Sect. 2.3.1) to remove nanoparticles not enclosed in the protein cavity, followed by size-exclusion chromatography (see Sect. 2.3.2) to isolate protein monomers. Protein concentration was determined using a Bradford assay (see Sect. 2.2.4).

To compare the effect of cobalt-doping on the magnetic properties of MF, un-doped MF was also synthesised following the procedure described above, with the exception that the iron precursor was added without the addition of cobalt sulphate.

2.6 Magnetoferritin Cationisation

For cationisation, *N*,*N*-dimethyl-1,3-propanediamine (DMPA) was coupled to aspartic and glutamic acid residues on the MF surface using *N*-(3-dimethylaminopropyl)-*N*′-ethylcarbodiimide hydrochloride (EDC) as described previously [31, 32]. Solutions of DMPA were adjusted to pH 7 using HCl, before drop-wise addition of MF or ApoF. After an equilibration period of 3 h, the coupling reaction was initiated by adding EDC and adjusting the pH to 5 using HCl. The solution was continuously stirred for up to 24 h and then filtered through a 220 nm syringe filter to remove any precipitates. The solution was extensively dialysed (Medicell dialysis tubing, 12–14 kDa MWCO) against 4 L of 50 mM phosphate buffer (pH 7) containing 50 mM NaCl to yield stable solutions of cat-MF and cationised ApoF (cat-ApoF).

2.7 Characterisation of MF and cat-MF

To determine the size of MF and cat-MF transmission electron microscopy (TEM) and dynamic light scattering (DLS) were used (see Sect. 2.4.2). Core sizes of MF nanoparticles were determined by image analysis of unstained samples, and the diameters of MF and ca-MF (core and protein shell) were determined by image analysis of stained samples, both by employing Image J (National Institute of Health, USA) software. The length of the scale bar in the recorded TEM images was converted to a pixel measure. A mean threshold was applied to clear the image from debris, which left discrete black and white images of the nanoparticles that was subjected to particle analysis, from 0—infinity size and 0–1 circularity. This yielded a list of particle sizes, for which the mean (m_x) and standard deviation (s_x) were calculated using Excel (Microsoft Co., USA) as follows:

$$m_x = \frac{1}{n}\sum_{i=1}^{n} x \tag{2.10}$$

and

$$s_x = \sqrt{\frac{1}{n-1}\sum_{i=1}^{n}(x - m_x)^2} \tag{2.11}$$

where n is the sample size and x the observed value.

The composition of the nanoparticle core was determined using inductively-coupled plasma optical emission spectrometry (ICP-OES) (see Sect. 2.4.5). Cationisation efficiency of covalent DMPA coupling to the MF surface was assessed with zeta potentiometry (see Sect. 2.4.3). For the assessment of the number of DMPA molecules conjugated to each ApoF subunit as a measure of cationisation efficiency, matrix-assisted laser desorption ionisation time-of-flight (MALDI-TOF) mass spectrometry was used (see Sect. 2.4.4). Characterisation of the magnetic properties of MF and cat-MF was performed using SQUID magnetometry and MRI (see Sects. 2.4.6 and 2.4.7).

2.8 Histology

2.8.1 *Fixing, Embedding, Sectioning and Dewaxing*

Theory

Tissues are prepared for histological staining and imaging through fixation, embedding and sectioning. Fixation preserves the tissue's micro-structure and embedding in a rigid matrix supports the tissue during thin sectioning, which in turn is required for microscopic observation. Tissues and stem cells used in this work were fixed with paraformaldehyde, which crosslinks proteins with methylene bridges, [33] and embedded in paraffin wax. Paraffin melts around 60 °C, such that the liquid can effectively infiltrate the tissue. Upon cooling, paraffin solidifies and tissues enclosed in this "block" can be sectioned using a microtome. Prior to staining, this wax needs to be removed by immersion into organic solvents, such as xylene.

Method

Cartilage constructs or hMSCs grown on fibronectin-coated polyglycolic acid (PGA) scaffolds (Biomedical Structures, USA) were fixed for 3 h in 4% (w/v) paraformaldehyde solution at room temperature. After that the samples were immersed in 100% ethanol and submitted to the Histology Services Unit (Medical Sciences, University of Bristol), where they were embedded in paraffin wax, cut into 5 or 10 μm sections and placed on Polysine microscope slides (VWR, UK).

Before staining, sections were de-waxed and re-hydrated by sequential immersion into xylene, 100% ethanol, 70% ethanol and finally dH_20, each for two minutes.

2.8.2 *Haematoxylin and Eosin Stain*

Theory

Haematoxylin and eosin (H&E) are frequently used to stain the cell's nucleus purple and the cytoplasm pink, respectively [34]. To produce a functional dye, haematoxylin is oxidized to haematin, which in turn forms a cationic complex with metal ions, such as aluminium (III). The metal cation mediates association of the dye molecule with acidic species, like nucleic acids through a metal-acid bond [35]. Eosin on the other hand binds to basic amino acid residues, thus staining proteins in the cytoplasm and extracellular matrix [36].

Method

De-waxed tissue sections on microscope slides were immersed in Haematoxylin QS solution (Vector Laboratories, UK) for two minutes and excess stain was removed by washing in dH_2O. They were then placed in a 10 mg mL^{-1} eosin Y solution for five minutes and the excess stain was removed by immersing the slides in 70%

ethanol, followed by 20 s in 100% ethanol. Slides were left to air dry and mounted on glass cover slips (VWR, UK) using DPX mounting medium.

2.8.3 Safranin O and Fast Green Stain

Theory
Safranin O is a cationic dye molecule, which binds electrostatically to anionic species in the tissue. In particular, anionic sulphate groups in proteoglycans form stoichiometric complexes with individual safranin O molecules [37]. Healthy cartilage tissue is rich in proteoglycans, such that safranin O staining can be used as a measure of proteoglycan distribution in the extracellular matrix of tissue engineered cartilage constructs [38] The cytoplasm of the cells in the tissue are often counterstained with Fast Green, which is an anionic dye that interacts with cationic amino acid residues in a similar fashion to eosin [39].

Method
De-waxed and re-hydrated tissue sections were immersed in a 1 mg mL^{-1} Fast Green solution for four minutes. The excess stain was removed by washing in 1% (v/v) acetic acid solution. Then, sections were placed in a 5 mg mL^{-1} Safranin O solution for six minutes. Excess stain was removed by dipping slides into 95% ethanol. Sections were de-hydrated in 100% ethanol for 20 s and then mounted onto glass cover slips with DPX mounting medium.

2.8.4 Prussian Blue and Nuclear Fast Red Staining

Theory
The Prussian Blue reaction is a histochemical method to stain iron deposits in cells and tissues, such as haemosiderin or ferritin [40, 41]. It is also widely used to visualise SPIONs internalised by cells [42]. The reaction involves a treatment with acidic solutions of ferrocyanides. Any ferric iron (Fe^{3+}) in cells/tissues combines with ferrocyanide and results in the formation of ferric ferrocyanide, a bright blue pigment referred to as Prussian Blue:

$$3\left[\mathrm{Fe}\,(\mathrm{CN})_6\right]^{4-} + 4\,\mathrm{Fe}^{3+} \rightarrow \mathrm{Fe}_4\left[\mathrm{Fe}\,(\mathrm{CN})_6\right]_3$$

Nuclear Fast Red is used as a red nuclear counterstain. Similar to haematoxylin, it is prepared in a solution containing excess aluminium ions, such that it becomes a cationic molecule that can bind to nucleic acids through a metal-acid bonds [43].

Method
De-waxed and re-hydrated sections were immersed in a 10% (w/v) potassium ferrocyanide solution mixed with an equal volume of 2 M hydrochloric acid for

20 min at room temperature. Slides were washed in dH_2O three times for five minutes. Cell nuclei were counter stained with a Nuclear Fast Red solution for five minutes and the excess stain was removed by rinsing with dH_2O. Sections were de-hydrated first in 70% then in 100% ethanol for 20 s and then mounted onto glass cover slips with DPX mounting medium.

2.8.5 *Immunostaining of Collagen Type II*

Theory

Immunohistochemistry makes use of antibodies to localise particular antigens in cells or tissues and stain those regions with high specificity [44]. However, during fixation of cells and tissues cross-linked proteins may mask these antigens. Therefore, they must first be retrieved and a common way of achieving this is to use proteolytic digestion. In the fixed cartilage tissue sections used here, proteoglycans and hyaluronic acid components of the extracellular matrix were digested with pronase and hyaluronidase, respectively. This exposed the collagen fibres present in the tissue, such that collagen type II epitopes (i.e. the regions of the antigen that the antibody binds to) could be targeted using an indirect immune-enzyme strategy (Fig. 2.18). In the first step, a primary antibody binds to collagen type II epitopes. Then, a biotinylated secondary antibody is added to bind to the primary antibody. Avidin is added to form strong electrostatic bonds with the biotin moiety of the secondary antibody. At the same time, avidin possesses binding sites for further biotinylated molecules, such as horseradish peroxidase. Therefore, a complex comprising avidin and biotinylated-horseradish peroxidase can bind to the secondary antibody. Horseradish peroxidase is an enzyme that can polymerise water

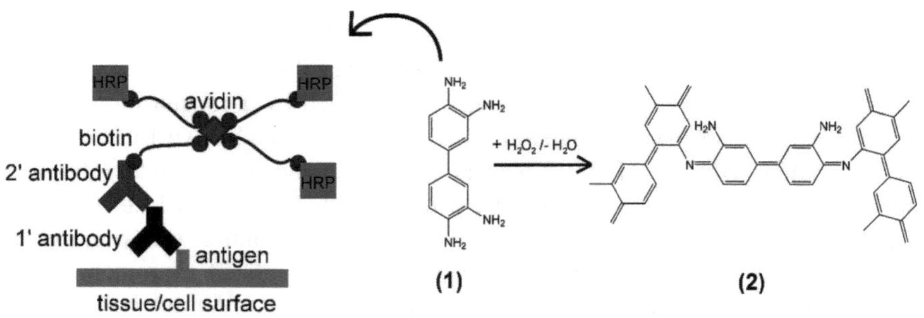

Fig. 2.18 Principle of the indirect immune-enzyme staining strategy. A primary (*1′*) antibody binds to the epitope in the tissue or on the cell surface. A biotinylated secondary (*2′*) antibody is added to bind to the *1′*. Avidin is added and forms electrostatic bonds with the biotin moiety of the 2′ antibody. A complex comprising avidin and biotinylated-horseradish peroxidase (HRP) can bind to the 2′ antibody. Horseradish peroxidase catalyses the polymerisation of water soluble 3,3-diaminobenzidine (DAB) tetrachloride (*1*) to an insoluble indamine polymer (*2*) through the reduction of hydrogen peroxide (H_2O_2)

soluble 3,3-diaminobenzidine (DAB) tetrachloride to an insoluble precipitate, [44, 45] which appears as a brown stain. Because the final step relies on peroxidase activity, any such innate activity in the tissue must be quenched to avoid unspecific staining. This is often achieved by incubation with hydrogen peroxide.

Method

De-waxed and re-hydrated sections were circled with a PAP pen (The Binding Site, UK) to create hydrophobic enclosures. 200 μL of a 10 mg mL^{-1} bovine testicular hyaluronidase solution was added to each section and the slides were incubated in a humidified chamber at 37 °C for 30 min. Slides were immersed in PBS for five minutes to remove excess enzyme. After that, 200 μL of a 2 mg mL^{-1} pronase solution (from *Streptomyces griseus*, Roche, Switzerland) was added to each section and the slides were incubated in a humidified chamber at 37 °C for 30 min. Slides were immersed in PBS for five minutes to remove excess enzyme. Then, 200 μL of 3% (v/v) hydrogen peroxide solution was added to each section, and they were incubated in a humidified chamber at room temperature for 5 min. Slides were immersed in PBS for five minutes to remove excess hydrogen peroxide. 200 μL of a blocking solution containing 30 mg mL^{-1} bovine serum albumin (BSA) solution was added and slides incubated in a humidified chamber at room temperature for 1 h. This was done to minimise unspecific binding of the primary antibody. Slides were washed by placing them three times for five minutes in fresh PBS.

Primary antibodies were prepared by dilution in **antibody diluent**, which consisted of 10 mg mL^{-1} BSA in Tris buffer (50 mM, pH 7.6) containing 150 mM NaCl and 0.05% (v/v) Tween 20. 70 μL of 1% (v/v) goat anti-type II collagen (Cambridge Biosciences, UK) or 5% (v/v) goat IgG isotype (Santa Cruz, USA) was added to each section and incubated in a humidified chamber at 4 °C overnight. Excess primary antibody solution was removed by a 15 min immersion in Tris buffer (50 mM, pH 7.6) containing 1 M NaCl, followed by two 15 min washes in Tris buffer. The secondary antibody, biotinylated Anti-Goat IgG (VECTASTAIN Elite ABC Kit, Goat IgG, Vector Laboratories, UK), was prepared as 0.5% solution in antibody diluent. 70 μL of the secondary antibody solution was added to each section and slides were incubated in a humidified chamber at room temperature for 1 h. Excess secondary antibody solution was removed by a 15 min immersion in Tris buffer (50 mM, pH 7.6) containing 1 M NaCl, followed by two 15 min washes in Tris buffer. 200 μL of ABC reagent containing the avidin and biotinylated horseradish peroxidase macromolecular complex (VECTASTAIN Elite ABC Kit) was added to the sections for 30 min. Excess reagent solution was removed by washing three times for 10 min in Tris buffer. 200 μL of diaminobenzidine tetrahydrochloride substrate (ImmPACT DAB Substrate Kit, Vector Laboratories, UK) was added to the sections for exactly ten minutes, and the excess substrate removed by rinsing the slides in dH_2O. Sections were de-hydrated by serial immersion into 70 and 100% ethanol for 2 min, and finally into xylene for 2 min to remove traces of the PAP pen. Slides were left to dry and then mounted on glass cover slips using DPX mounting medium.

2.9 Immunohistochemical Methods

2.9.1 Cartilage Digestion

Theory

Release and separation of extracellular matrix proteins, such as collagen and proteoglycans, from the PGA scaffold was achieved through digestion of engineered cartilage constructs using trypsin. Trypsin is a serine protease that cleaves C-terminal amide bonds of arginine and lysine residues [46]. An array of inhibitors was added during the digestion process to prevent proteolysis of the collagen type II epitope by other enzymes. Chymotrypsin activity was inhibited with tosyl phenylalanyl chloromethyl ketone (TPCK), [47] cysteine peptidases were inhibited using iodoacetamide, [48] aspartyl proteases were inhibited with pepstatin A, [49] and metalloproteinases were inhibited with EDTA [50]. After digestion, trypsin can be inactivated by heat-induced denaturation.

Prior to digestion, any residual liquid needs to be removed from the constructs to determine their dry weight. To this end, lyophilisation (or freeze-drying) was employed. This is a method for dehydrating aqueous samples without applying heat. Instead, samples are frozen in a low pressure chamber, which causes the sublimation of solvent molecules.

Method

Cartilage constructs were placed in 2 mL tubes (Eppendorf, UK) and lyophilised at −50 °C and low pressure (< 0.05 mBar) overnight using a ModulyoD freeze-drier (Fisher Scientific, UK). The dry constructs were weighed to find the dry mass of the whole cartilage construct. Dried constructs were digested in 50–350 μL of a 2 mg mL^{-1} digestion solution containing TPCK-treated trypsin supplemented with 200 mM iodoacetamide, 200 mM EDTA and 2 mg mL^{-1} pepstatin A at 37 °C in a Vortemp 56 shaking incubator (Labnet, USA) overnight. A further 50–350 μL of the digestion solution was added and then tubes were incubated for 2 h at 65 °C in a shaking incubator and vortex-mixed at 10 min intervals during the first hour. Samples were boiled for 15 min at 100 °C to inactivate trypsin and then centrifuged for two minutes at 12300 × g. The supernatant containing the digested cartilage matrix components was removed and the undigested scaffold freeze-dried and weighed again to find the mass of the scaffold. The dry weight of the extracellular matrix was calculated by subtracting the dry weight of the undigested scaffold from the dry weight of the whole cartilage construct.

2.9.2 Collagen Type II Assay

Theory

Collagen type II is an integral part of the extracellular matrix component in cartilage [51]. Collagen type II was quantified by performing an inhibition enzyme-linked

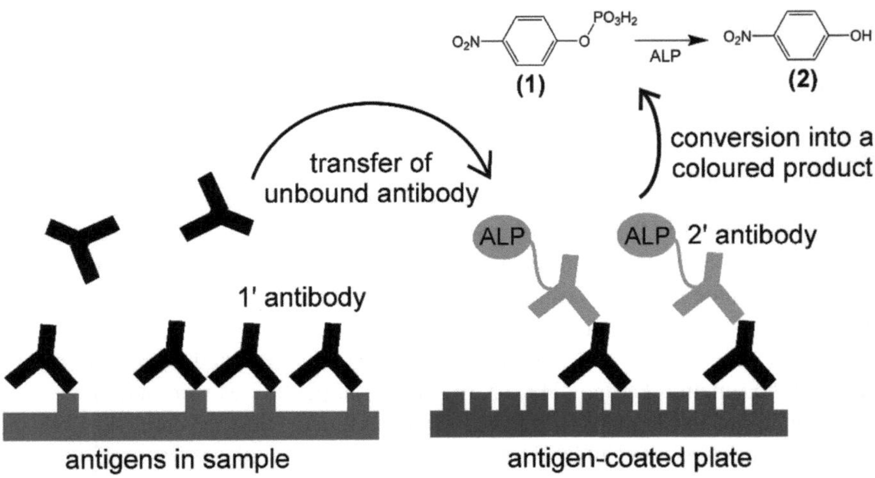

Fig. 2.19 Principle of the inhibition ELISA to quantify collagen type II. An excess of primary antibody is added to bind to antigens in the sample. The supernatant containing unbound primary antibody is transferred to a plate coated with collagen type II and binds to the antigens. An alkaline phosphatase (ALP)-conjugated secondary antibody binds to the immobilised primary antibody and catalyses the hydrolysis of colourless *p*-nitrophenyl phosphate (*1*) to yellow *p*-nitrophenol (*2*)

immunosorbent assay (ELISA). This assay was developed by Hollander et al, [52] who used the peptide CB11B (GKVGPSGA[Hyp]GEDGR[Hyp]GP[Hyp]GP) as a marker of denatured type II collagen. In the assay, a mouse antibody against type II collagen is incubated in excess with a range of CB11B standards and digested cartilage solutions. A set volume of each solution is transferred to a new plate with CB11B-coated wells. Any primary antibody not bound to collagen or peptide is available to bind the coated wells. Therefore, the more type II collagen present in the initial solution, the fewer antibody molecules bind to the wells of the second plate. A secondary antibody conjugated to alkaline phosphatase (ALP) is used to bind to the primary antibody. Alkaline phosphatase catalyses the conversion of colourless *p*-nitrophenyl phosphate substrate into a yellow *p*-nitrophenol product. This colour change can be quantified through absorbance measurements using a spectrophotometer. The absorbance signal from a series of peptide standards can be used to build a standard curve to estimate the quantity of type II collagen present in the digested cartilage matrix (Fig. 2.19).

Method

Collagen type II coated plates were prepared by coating the inner wells of a Costar Immulon-2 high binding 384 well plate (Appleton Woods) with 40 μL of heat-denatured type II collagen solution (Bioiberica, Spain). Plates were wrapped in cling film and left to incubate at 4 °C for three days. Plates were washed three times in PBS containing 0.1% (v/v) Tween 20 and left to dry. 50 μL of a 10 mg mL^{-1} BSA blocking solution was added to each well and incubated for 30 min at room temperature. Plates were washed in PBS containing 0.1% (v/v) Tween, dried at 37 °C for 20 min and stored wrapped in cling film at 4 °C until use.

80 μL of a 10 mg mL^{-1} BSA blocking solution was added to the inner wells of a fresh Costar 384 well plate and incubated for 30 min at room temperature. Plates were washed in PBS containing 0.1% (v/v) Tween and left to air dry. Samples along with calibration standards of collagen II (Collagen Type II fragment, CB11B, Pepceuticals, UK; concentrations ranging from 0.5–6 μg mL^{-1}) were plated out in triplicate and incubated with a 0.17% (v/v) mouse-derived collagen type II antibody (COL2–3/4) at 37 °C overnight. After that, samples were transferred into the prepared collagen type II-coated plates for exactly 30 min at room temperature. The plates were then washed three times, and 10 μL of a 0.1% (v/v) solution of an ALP-conjugated goat anti-mouse secondary antibody (Cambridge Biosciences, UK) prepared in a 10 mg mL^{-1} BSA blocking solution was added to each well. The plates were sealed and incubated for 2 h at 37 °C. A disodium *p*-nitrophenyl phosphate tablet was dissolved in 10 mL of diethanolamine buffer, yielding a solution of approximately 0.5 mg mL^{-1}. After incubation with the secondary antibody, the plates were washed, 10 μL of the *p*-nitrophenyl solution was added to all wells and the plates left to incubate for 20 min at 37 °C. Then absorbance was measured at 405 nm using UV/Visible spectroscopy. The amount of collagen II was calculated from the mean absorbance of the CB11B peptide standards and expressed as a percentage of the extracellular matrix dry weight. The percentage of collagen II in cat-MF treated and untreated cartilage constructs was compared using the Mann-Whitney U-test in IBM SPSS Statistics version 21. A non-parametric test was chosen because of the small sample size, which made it difficult to confirm normality for the data sets. The Mann-Whitney U-test was employed because it is a useful test to compare differences between two treatment groups.

2.9.3 Glycosaminoglycan Assay

Theory

Glycosaminoglycans (GAG) are long unbranched polysaccharides containing repeating disaccharide units, and chondroitin sulphate is an abundant GAG in cartilage [53]. Chondroitin sulphate is an anionic molecule that forms an electrostatically bonded complex with the cationic dye 1,9-dimethylene blue (DMMB) [54]. Therefore, GAG content in engineered cartilage constructs can be determined using a DMMB-based assay. The binding of chondroitin sulphate causes a shift in the absorption maximum of DMMB, from 649 nm for the free molecule to 525 nm for the complex [55].

Method

20 μL of the digested cartilage construct matrix was added to a 96 well plate, alongside a series of chondroitin sulphate standards (5–50 μg mL^{-1}). 250 μL of a 160 μg mL^{-1} dimethylmethylene blue solution was added to all wells and the

absorbance measured immediately at 530 nm using UV/Visible spectroscopy. The mean absorbance of the chondroitin sulphate standards were used to plot a standard curve and estimate the GAG concentration in the digested cartilage constructs, which was expressed as a percentage of the construct's dry weight. The percentage of GAG in cat-MF treated and untreated cartilage constructs was compared using the Mann-Whitney U-test in IBM SPSS Statistics version 21 (see Sect. 2.9.2 for rationale behind statistical test).

2.10 Mesenchymal Stem Cell Culture

2.10.1 Culture Conditions

Human mesenchymal stem cells (hMSC) were harvested from the proximal femur bone marrow of osteoarthritic patients undergoing total hip replacement surgery, in full accordance with Bristol Southmead Hospital Research Ethics Committee guidelines (reference #078/01) and after patient consent was obtained. All cell culture was performed in Hera Safe class II laminar flow cabinets (Kendro, Germany) and incubated in humidified Hera Cell 150 incubators (Kendro, Germany) at 37 °C and 5% carbon dioxide atmosphere. The cells were cultured as monolayers using 175 cm^2 flasks (Appleton Woods, UK) and 20 mL of culture medium, which was replenished every 3–4 days (Fig. 2.20). **Culture medium** consisted of Dulbecco's Modified Eagle's Medium (DMEM), containing 1000 mg L^{-1} glucose, 10% (v/v) foetal bovine serum (FBS), 1% (v/v) penicillin/streptomycin solution, 1% (v/v) glutamax solution (Life Technologies) and 5 ng mL^{-1} freshly supplemented human fibroblast growth factor (PeproTech).

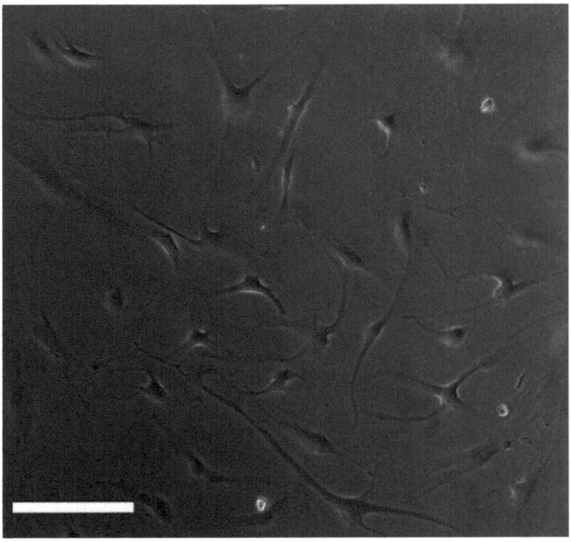

Fig. 2.20 Light microscopic image of human mesenchymal stem cells in culture (passage 6). Magnification 10×. Scale bar 100 μm

The hMSCs were inspected before each media change using bright field microscopy, and harvested and re-seeded (passaged) once they reached confluency. hMSCs were used for experiments up to passage 6.

2.10.2 Harvesting, Counting and Passaging hMSCs

Cells were harvested either for passage, cryopreservation or in preparation for an experiment. Culture medium was aspirated and cells washed with 20 mL PBS. 5 mL of trypsin/EDTA solution was added to the flask and incubated at 37 °C for five minutes. Cell detachment was confirmed using bright field microscopy, before addition of 10 mL of culture medium. The cell suspension was removed into 15 mL centrifuge tubes (Appleton Woods, UK) and centrifuged for five minutes at $524 \times g$. The supernatant was removed and the cell pellet re-suspended in culture medium. A 10 μL sample of the cell suspension was placed in an AC1000 improved Neubauer haemocytometer (Hawksley, UK). For passage, hMSCs were re-seeded at a low cell density in 175 cm^2 flasks in 20 mL culture medium.

2.10.3 Freezing and Thawing hMSCs

Cells were harvested, counted, and samples containing 10^6 hMSCs were centrifuged and the supernatant removed. The cell pellet was re-suspended in 1 mL of freezing medium, which consisted of two parts 10% (v/v) dimethyl sulfoxide in FBS, and one part of culture medium. The cell suspension was transferred into 1.5 mL cryovials (Fisher Scientific, UK) and placed in a Mr. Frosty freezing container (Fisher Scientific, UK) overnight at −80 °C. After that they were transferred to a Biorack 6000 refrigerator (Statebourne Cryogenics, UK), where they were stored in the vapour phase of liquid nitrogen (approximately −190 °C).

Cryopreserved hMSCs were thawed in a 37 °C water bath, transferred immediately into a 175 cm^2 flask containing 20 mL of culture medium and cultured as described in Sect. 2.10.1.

2.11 Stem Cell Labelling with cat-MF

This section describes the procedures carried out to investigate magnetic labelling of hMSCs with cat-MF. To assess labelling rapidity and efficiency, MACS was used. First, the characterisation of the MACS magnet is summarised, followed by the description of the MACS experiments. Then, the procedures to image hMSCs labelled with cat-MF using MRI, TEM and light microscopy are reported. Finally, an experiment to elucidate the labelling mechanism is described.

2.11.1 Characterisation of the MACS Magnet and Column

The magnetic field of a miniMACS separator magnet (Miltenyi Biotec) (Fig. 2.21a) was measured with a Gaussmeter and a transverse Hall Effect probe (Hall Effect Gaussmeter, Model 5170 F.W. Bell, Sypris Test & Measurement). The probe measured magnetic flux density B perpendicular to a sensing area of 0.38 mm in diameter, located 0.85 mm from the end of the rectangular probe rod (Fig. 2.21b).

The probe was placed in the MACS column cavity and B was recorded every millimetre along the entire length of the magnet from magnetic North to South (Fig. 2.22a). The magnetic field gradient was calculated from the linear portion of a plot of B as a function of position inside the cavity and found to be 57 mT/mm over the first and last 10 mm. In the middle of the cavity the field remained constant at approximately 500 mT over 10 mm (Fig. 2.22b).

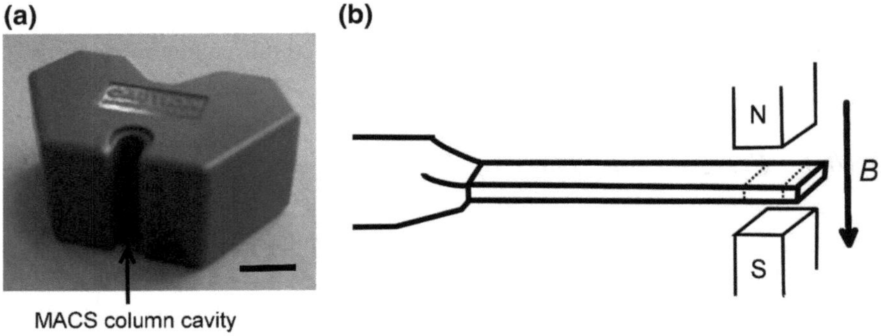

Fig. 2.21 **a** Photograph of the MACS magnet. *Scale bar* 1 cm. **b** Transverse Hall probe measuring magnetic flux density B perpendicular to a sensing area (*dashed outline*)

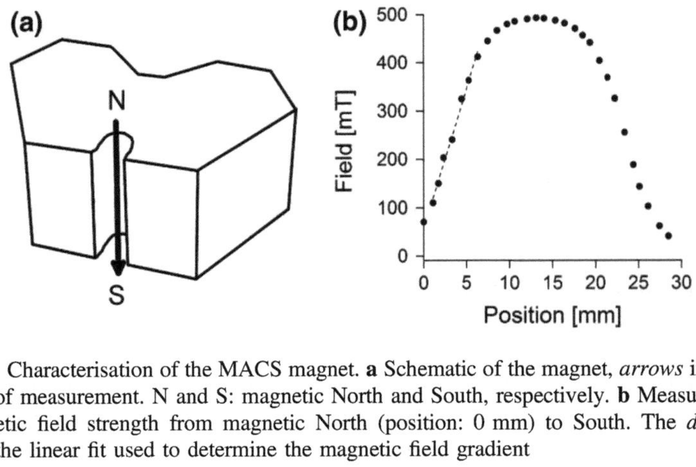

Fig. 2.22 Characterisation of the MACS magnet. **a** Schematic of the magnet, *arrows* indicate the direction of measurement. N and S: magnetic North and South, respectively. **b** Measurements of the magnetic field strength from magnetic North (position: 0 mm) to South. The *dashed line* indicates the linear fit used to determine the magnetic field gradient

The ferromagnetic matrix of the MACS columns used with this magnet was 20 mm long and extended between approximately 5 - 25 mm along the area measured in Fig. 2.22b. The diameter of the column was 4 mm and it fitted tightly into the width of the MACS column cavity. According to the manufacturer, the column matrix is composed of ferromagnetic beads with a diameter of 250 to 500 μm, coated in a polymeric material to protect cells from damage, and the space between the spheres is 100–200 μm. The flow rate in the column was measured to be approximately 0.5 mL min^{-1}.

2.11.2 Magnetic Stem Cell Labelling: Time and Concentration Dependence

To assess the rapidity of magnetic labelling with cat-MF, 150000 hMSCs from three different patients were seeded in culture medium into 25 cm^2 tissue culture flasks and left to adhere over night. Cells were washed with PBS and exposed to 1 mL of 0.5 μM cat-MF in PBS for 1 min to 6 h. After labelling, cells were washed with PBS and harvested from the flasks using trypsin/EDTA. Cell pellets were re-suspended in 0.5 mL magnetic separation buffer (0.5% (w/v) FBS and 2 mM EDTA in PBS) and magnetic-activated cell separation (MACS) was performed using MACS MS columns (Miltenyi BIOTEC) with a capacity to immobilise up to 10^7 magnetically labelled cells. First 0.5 mL of the sample in MACS buffer was loaded into the reservoir of the column. Once the reservoir was empty, 0.5 mL of MACS buffer was added three times (total volume 1.5 mL) to elute cells that were not magnetised sufficiently to be retained in the column. After that, the column was detached from the magnet, 1 mL of MACS buffer was added and pushed thought the column using a plunger. This eluted the magnetised cells from the column matrix. Cell numbers in the magnetic and non-magnetic cell fractions were determined using an Improved Neubauer haemocytometer (see Sect. 2.10.2). To assess magnetisation efficiency, the fraction of magnetised cells, $M(\%)$, was determined as follows:

$$M(\%) = 100 \cdot \frac{n(M)}{n(M+NM)} \tag{2.12}$$

where $n(M)$ is the number of cells in the magnetic fraction and $n(M + NM)$ is the sum of the number of cells in the magnetic and non-magnetic fractions. Cells in the magnetic fractions were digested with 50% (v/v) nitric acid before iron content was determined using ICP-OES. The quantity of iron measured was normalised to the number of cells in the analysed fraction. The saturation value for cellular iron, Fe_{max}, and the time constant for iron uptake, k, were determined by fitting an exponential function to the experimental data over the 6 h time course:

$$Fe\,(t) = Fe_{\max}\left[1 - \exp\left(-\frac{t}{k}\right)\right] \tag{2.13}$$

where $Fe(t)$ is the cellular iron content at a particular time point and t is the incubation time in hours.

To assess the effect of cat-MF concentration on labelling efficiency, 150000 hMSCs from three different patients were seeded in culture medium into 25 cm^2 tissue culture flasks and left to adhere over night. Cells were washed with PBS and exposed to 1 mL of cat-MF concentrations ranging from 0.01 to 2 μM in PBS for 30 min. After labelling, cells were washed with PBS and harvested from the flasks. MACS and ICP-OES were performed as described above.

Finally, to confirm the effect of cationisation on labelling efficiency, 150000 hMSCs from three different patients were seeded in culture medium into 25 cm^2 tissue culture flasks and left to adhere over night. Cells were washed with PBS and exposed to 1 mL of MF concentrations ranging from 0.5 to 3 μM in PBS for 30 min. After labelling, cells were washed with PBS and harvested from the flasks. MACS and ICP-OES were performed as described above.

To calculate the minimum number of cat-MF particles needed to magnetise hMSCs sufficiently for retention in the MACS column, the magnitude of the magnetic force, F_m, acting on the magnetically labelled cells needs to compensate the drag force, F_d, experienced during the flow through the column. The magnetic force on a labelled cell can be approximated as:

$$F_m = N \cdot \mu \cdot \nabla B \tag{2.14}$$

where N is number of magnetic particles on the cell, μ is the magnetic moment of an individual magnetic nanoparticle (which can be determined from the Langevin function, see Eq. 2.5), and ∇B is the magnetic field gradient. The drag force can be described by:

$$F_d = 6 \cdot \pi \cdot \eta \cdot r_{cell} \cdot v \tag{2.15}$$

where η is the viscosity of the medium in which the cells are suspended (predominantly water), r_{cell} is the radius of the cell and v is the velocity of the cell as it's passing through the column.

2.11.3 MRI Imaging of Labelled hMSCs

800000 hMSCs were seeded in culture medium in 75 cm^2 tissue culture flasks and left to adhere over night. The cells were labelled with 3 mL of 0.5 μM of MF or cat-MF for 30 min. The MF or cat-MF supernatant was removed and cells were washed with PBS, harvested with trypsin/EDTA and counted. 750000 labelled cells were suspended in 200 μL of PBS and transferred into tubes of a PCR plate embedded in a 1.5% (w/v) carrageenan gel and left to settle by gravity for 4 h prior to MRI imaging. T1 and T2 were determined as described in Sect. 2.4.7. Additionally, a gradient echo sequence was used with TR = 100 ms, TE = 10 ms, and a flip angle of 30° to image the labelled cells.

2.11.4 TEM Imaging of Labelled hMSCs

50000 hMSCs were seeded in culture medium into a 6 well plate and left to adhere over night. The cells were washed with PBS and exposed to 1 mL of 0.5 μM cat-MF for 30 min. The cat-MF supernatant was removed and the cells were washed with PBS. The cells were either immediately fixed and stained for TEM imaging or cultured for one week before fixing and staining (see Sect. 2.4.1).

2.11.5 Prussian Blue Staining of Labelled hMSCs

1000000 hMSCs were seeded in culture medium into 75 cm^2 tissue culture flasks and left to adhere overnight. The cells were exposed to 3 mL of 0.5 μM of cat-MF for 30 min. The cat-MF supernatant was removed, cells were washed with PBS and left in culture medium for 24 h. After that, hMSCs were harvested with trypsin/EDTA and counted. 300000 cells were suspended in 30 μL of culture medium and loaded onto fibronectin-coated PGA tissue engineering scaffolds of 5 mm diameter and 2 mm thickness (Biomedical Structures, USA) and placed in agarose coated wells of a 24 well plate. Cells were left to attach to the scaffold over night, and then cultured for 6 h in culture medium. After that, the medium was removed and scaffolds fixed in 4% (w/v) paraformaldehyde over night at room temperature, dehydrated in 70% ethanol for 2 h and then submitted to the histology service of the School of Cellular and Molecular Medicine at the University of Bristol. Histological sections of 10 μm thickness were stained with Prussian Blue and Nuclear Fast Red (see Sect. 2.8.4) and observed using an inverted light microscope at 40 × magnification (Leica DM IRB).

2.11.6 Long Term Retention of Magnetisation

To investigate the long term retention of magnetisation following labelling with cat-MF, 150000 cells were seeded in culture medium into 75 cm^2 tissue culture flasks and left to adhere over night. The cells were washed with PBS and exposed to

3 mL of 0.5 μM cat-MF for 30 min. The cat-MF supernatant was removed, the cells washed with PBS and then cultured for up to five weeks. Magnetisation and iron content of the cells were analysed at weekly intervals using MACS and ICP-OES.

To assess the long term retention of cat-MF for MR imaging, 800000 cells were seeded into 75 cm^2 flasks, left to adhere over night, and then labelled with 3 mL of 0.5 μM of cat-MF for 30 min. The cat-MF supernatant was removed, the cells were washed with PBS, and cultured in DMEM for either four days or one week before being harvested. 750000 labelled hMSCs were suspended in 200 μL of PBS and transferred into tubes of a PCR plate embedded in a 1.5% (w/v) carrageenan gel and left to settle by gravity for 4 h prior to MRI imaging using a gradient echo sequence with TR = 100 ms, TE = 10 ms, and a flip angle of 30°. T1 and T2 were also determined as described in Sect. 2.4.7.

2.11.7 Elucidating the Labelling Mechanism

To evaluate the contribution of active internalisation processes on cellular iron content 150000 hMSCs were seeded in culture medium into 25 cm^2 tissue culture flasks and left to adhere over night. hMSCs were exposed to 1 mL of 0.5 μM cat-MF for 5 or 30 min at 37 °C or 4 °C. Prior to incubation with cat-MF at 4 °C, cells were pre-cooled at 4 °C for 30 min in DMEM supplemented with 20 mM HEPES buffer. After each incubation period, the cat-MF supernatant was removed and the cells were washed, harvested and analysed using ICP-OES.

To test the hypothesis that anionic proteoglycans in the glycocalyx facilitate cat-MF uptake 150000 cells were cultured for four days in 25 cm^2 tissue culture flasks using DMEM supplemented with 80 mM sodium chlorate ($NaClO_3$), because $NaClO_3$ inhibits the synthesis of anionic functional groups on proteoglycans [56]. hMSCs were labelled with 1 mL of 0.5 μM cat-MF for 5 or 30 min at 37 °C before the cat-MF supernatant was removed and the cells were washed, harvested and analysed using ICP-OES.

Statistical analysis was undertaken for each time point separately (5 and 30 min). The cellular iron content values in hMSCs labelled under different conditions were compared using the Kruskall-Wallis test in IBM SPSS Statistics version 21. A non-parametric test was chosen because of the small sample size (n = 3), which made it difficult to confirm normality for the data sets. The Kruskal-Wallis test was used because more than two groups were compared. The statistics program also conducted pair-wise analysis as part of the procedure, which identified differences between groups.

2.12 Assessing Cell Viability After cat-MF Labelling

Theory

A common method to probe cell viability is the assessment of cellular metabolic activity. Cell metabolism results in the production of adenosine tri phosphate (ATP), which is the "energy currency" that is required for many processes in the living cell. A so-called electron transport chain is used to synthesis ATP, and thus, assessing the function of this vital process gives an indication of cell viability. Many commercially available biochemical assays probe metabolic activity by incubating cells with a dye that enters the cell and can be reduced by electrons from the electron transport chain. The formation of the reduced product can be quantified spectrophotometrically by measuring absorbance or fluorescence at a characteristic wavelength, and absorbance or fluorescence intensity are directly proportional to the number of metabolically active cells.

Among the most widely used dyes are tetrazolium salts, such as 3-(4,5-dimethylthiazol-2-yl)-5-(3-carboxymethoxyphenyl)-2-(4-sulfophenyl)-2H-tetrazolium (MTS) or 3-(4,5-dimethylthiazol-2-yl)-2,5-diphenyltetrazolium bromide (MTT). The dye is formulated into an assay solution that can be dissolved in culture medium. When the assay solution is added to the cells, the dye can access the intracellular space where it is reduced to formazan. This process is thought to be mediated by enzymes of the electron transport chain that generate reducing agents, such as nicotinamide adenine dinucleotide hydride (NADH). NADH can transfer its electrons to an electron transfer reagent (ETR), which is usually incorporated in the assay solution. Thus, the ETR is reduced and subsequently interacts with the tetrazolium compound, reducing it to a deeply coloured formazan product (Fig. 2.23a) [57]. Formazan has an absorbance maximum at 490 nm and can therefore be quantified by measuring absorbance at this wavelength. The measured absorbance values can then be compared between cells treated with a test compound and untreated cells to assess the impact of the test compound on cell viability.

Similarly, resazurin is a molecule that can be taken up by many cell types, and is used in the so-called Alamar Blue assay. If the cells are in a proliferative state, the reductive intracellular environment will lead to the reduction of resazurin to fluorescent resorufin using NADH (Fig. 2.23b) [58]. Fluorescence intensity can be measured upon excitation of the fluorescent compound at 570 nm and measuring emission at 590 nm. Again, the measured fluorescence intensity values can be compared between cells treated with a test compound and untreated cells to assess the impact of the test compound on cell proliferation. Both the MTS and the Alamar Blue assay measure similar endpoints, but the advantage of the Alamar Blue assay is that cells can be incubated with resazurin over prolonged periods of time, during which the formation of the resorufin product can be measured repeatedly. Therefore, it can be investigated whether adverse effects are transient or persistent. Both assays are sensitive to cell number and incubation time with the dye, therefore, both parameters were optimised for the hMSCs used here.

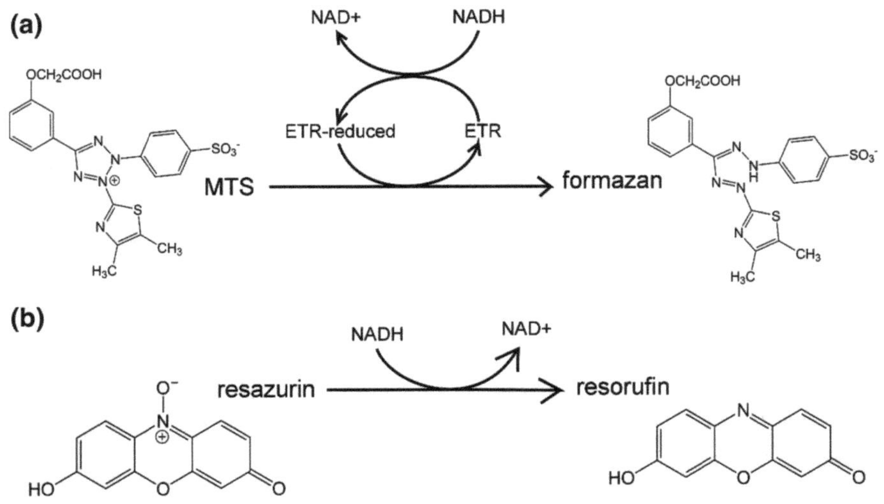

Fig. 2.23 Chemistry of viability assays. **a** The reduction of MTS to formazan through electron transfer reactions inside viable cells. An electron transfer reagent (ETR) is often added to the assay formulation, which acts as an intermediate electron acceptor and results in the formation of a soluble formazan compound. **b** Similarly, reduction of resazurin to fluorescent resorufin happens in the reductive environment of viable cells, mediated by nicotinamide adenine dinucleotide hydride (NADH)

2.12.1 Optimisation of the MTS and Alamar Blue Assay

Formation of the coloured products that are measured in the MTS and Alamar Blue assay is highly dependent on the number of cells present in the sample. Therefore, an appropriate number of cells needs to be chosen to ensure the validity of these assays. The ideal cell number can be determined by performing the assays on a variety of cell numbers and determining which cell numbers produce a signal that lies within the linear range of the assay.

A range of hMSC numbers between 1000 and 10000 was seeded in triplicate into 96 well plates and left to adhere over night. Culture medium was removed, cells washed with PBS and either incubated for 1 h at 37 °C with 100 μL of phenol-free culture medium containing 20% (v/v) of MTS solution (CellTiter 96, Promega), or for 4, 24 and 48 h at 37 °C with 100 μL of culture medium containing 10% (v/v) Alamar Blue solution (Serotec). Formation of reduced dye products was assessed by either measuring absorbance at 490 nm (MTS assay) or fluorescence intensity at excitation 570 nm and emission 590 nm (Alamar Blue assay) using a spectrophotometer.

For the MTS assay, a linear response was observed when the MTS reagent was incubated for 1 h with 1000 to 10000 hMSCs (r^2 = 0.98; Fig. 2.24a). Similarly, the fluorescence intensity of Alamar Blue increased linearly with cell number after 4 and 24 h incubation (r^2 = 0.98 and 0.99, respectively). After 48 h, the fluorescence

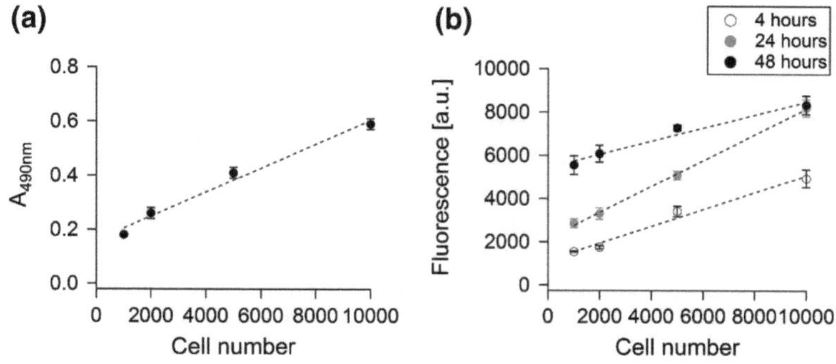

Fig. 2.24 Optimisation of cell viability assays. **a** Absorbance in the cell culture supernatants of 1000 to 10000 hMSCs exposed to MTS reagent for 1 h at 37 °C. **b** Fluorescence intensity (excitation 570 nm, emission 590 nm) in wells containing 1000 to 10000 hMSCs incubated with Alamar Blue reagent for 4, 24 and 48 h at 37 °C

intensity value recorded for 10000 cells was the same as the value recorded after 24 h (Fig. 2.24b). Therefore, it was concluded that a cell number below 10000 had to be chosen to obtain reliable results over a 48 h test period.

2.12.2 *MTS and Alamar Blue Assay After Exposure to MF and cat-MF*

For the **MTS assay**, 5000 cells from three different patients were seeded in culture medium in quadruplicate into wells of a 96 well plate, left to adhere over night, and exposed for 30 min to 125 μL of MF or cat-MF at concentrations ranging from 0.01 to 3 μM, or PBS only (untreated control). The supernatant was removed, cells washed with PBS and 100 μL culture medium was added to each well. After 24 h, culture medium was removed, cells washed with PBS and incubated for 1 h at 37 ° C with phenol-free culture medium containing 20% (v/v) of MTS solution. Formation of the reduced formazan product was measured at 490 nm using a spectrophotometer, and cell viability values of MF or cat-MF exposed cells (MF_{A490}) were normalised with respect to the untreated control (UTC_{A490}):

$$\%viability = \frac{MF_{A490}}{UTC_{A490}} * 100 \qquad (2.16)$$

For the **Alamar Blue assay**, 2000 cells from three different patients were seeded in culture medium in quadruplicate into wells of a 96 well plate, left to adhere over night and exposed for 30 min to 125 μL of MF or cat-MF at concentrations ranging from 0.01 to 3 μM, or PBS only (untreated control). The supernatant was removed, cells washed with PBS and 100 μL of culture medium containing 10% (v/v) Alamar Blue solution was added to each well. After 4, 24 and 48 h, fluorescence intensity was measured at excitation 570 nm and emission 590 nm using a spectrophotometer. Cell viability values were normalised with respect to the untreated control as described in Eq. 2.16.

For **statistical analysis** of results obtained from the MTS and Alamar Blue assays, data were presented as mean ± standard deviation of three biological repeats (total $n = 12$). Statistical analysis was performed using IBM SPSS Statistics version 21. The means of groups treated with MF, cat-MF, and PBS (untreated control) were compared using Two-Way Analysis of Variance (ANOVA), in which effects of dose as well as surface functionalisation (unfunctionalised vs. cationised) were investigated, followed by a one-sided Dunnett t post hoc test where appropriate. Two-Way-ANOVA was chosen because the data was found to be normally distributed and the variances were equal. ANOVA is a powerful test to analyse differences between more than two treatment groups, and Two-Way ANOVA can be used when the effect of two independent variables is to be explored (in this case dose and surface functionalisation).

2.13 Assessing Cytotoxicity After cat-MF Labelling

Theory

The outer cell membrane is a vital structure because it maintains a particular intracellular environment, and shields it from uncontrolled entrance of external fluids, molecules and organisms. Disruption of the cell membrane impacts on normal cell functions and often leads to cell death. Therefore, assessing membrane integrity after nanoparticle exposure gives an indication of potential cytotoxic effects that are mediated through disruption of the cell membrane. One way of probing if the cell membrane is still intact, is the detection of lactate dehydrogenase (LDH) in the cell culture medium. LDH is normally confined to the intracellular space and is released into the surrounding medium if the cell membrane is disrupted. Extracellular LDH can be detected in the medium using a colourimetric assay through a coupled enzymatic reaction in which LDH catalyses the conversion of lactate to pyruvate via NAD^+ reduction to NADH. The enzyme diaphorase is also added, which uses NADH to reduce the tetrazolium salt 2-(4-iodophenyl)-3-(4-nitrophenyl)-5-phenyl-2*H*-tetrazolium (INT) to a red formazan product that can be measured spectrophotometrically at 490 nm (Fig. 2.25) [59]. The amount of formazan is directly proportional to the amount of LDH released into the medium, and thus indicative of cytotoxicity.

The assay is sensitive to the number of cells, as well as to the presence of serum in the culture medium, which can give false positives. The manufacturer recommends the use of no more that 5% (v/v/) foetal bovine serum (FBS) in the culture medium.

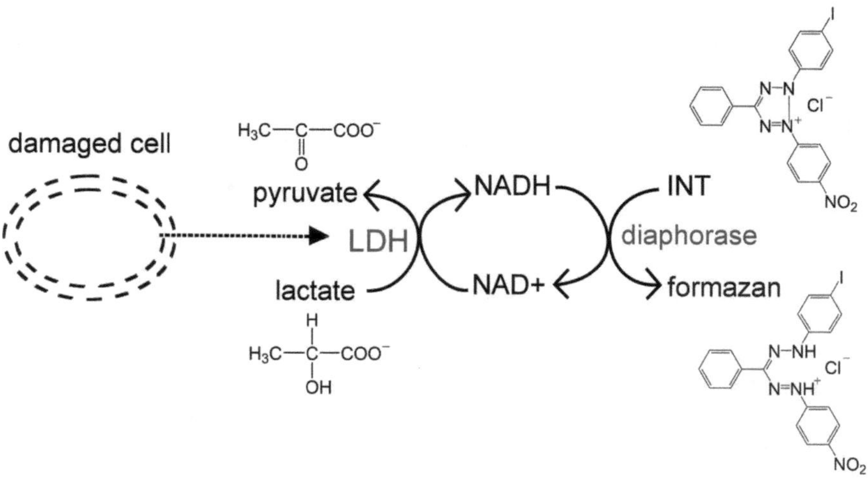

Fig. 2.25 Chemistry underlying the LDH assay. Damaged cell membrane leads to the leakage of LDH. Extracellular LDH is detected through a coupled enzymatic reaction in which LDH catalyses the conversion of lactate to pyruvate via NAD^+ reduction to NADH. Diaphorase uses NADH to reduce a tetrazolium salt (INT) to a *red* formazan product

2.13.1 Optimisation of the LDH Assay

Formation of the coloured product that is measured in the LDH assay is highly dependent on the number of cells present in the sample. Therefore, an appropriate number of cells needs to be chosen to ensure the validity of the assay. The ideal cell number can be determined by performing the assays on a variety of cell numbers and determining which cell numbers produce a signal that lies within the linear range of the assay. Because the signal is generated from intracellular LDH leaked into the culture medium, cells membranes were disrupted using a surfactant-based solution (lysis buffer) to achieve maximum LDH release.

1000 to 10000 hMSCs were seeded into wells of a 96 well plate in triplicate, and incubated over night in 100 μL culture medium containing 5% of FBS, and culture medium containing 5% of FBS and 10% (v/v) of a lysis buffer supplied by the manufacturer (LDH Cytotoxicity Assay Kit, Pierce). 50 μL samples of the medium supernatant were transferred to a separate 96 well plate, where they were mixed with an equal volume of "reaction mixture" supplied by the manufacturer (although the exact composition of this mixture is not disclosed by the supplier, it likely contains all the substrates needed for LDH detection, namely lactate, NAD, INT and diaphorase). After a 30 min incubation period at room temperature, 50 μL of a "stop solution" was added. Formation of the red formazan product was assessed by measuring absorbance at 490 nm using a spectrophotometer (SpectraMax M5e, Molecular Devices).

Incubation of the LDH reagent with cell culture medium from lysed hMSCs showed a linear increase in the absorbance at 490 nm over the whole range of cell

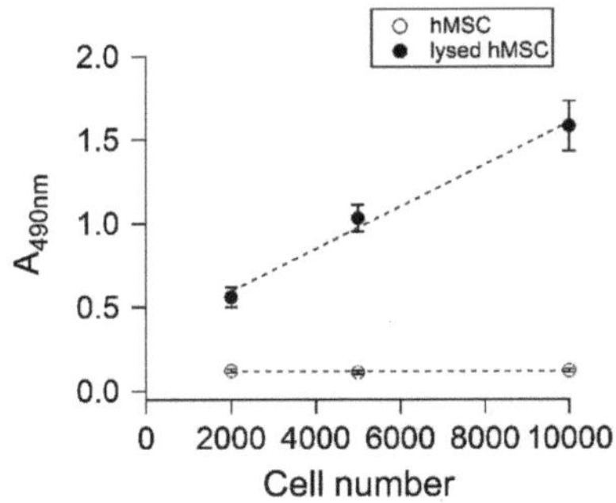

Fig. 2.26 Absorbance in cell culture supernatants of 2000 to 10000 untreated or lysed hMSCs incubated for 30 min with LDH reagent at room temperature

numbers ($r^2 = 0.99$, Fig. 2.26). Medium supernatants taken from untreated hMSCs and incubated with the LDH reagent did not result in an increase in absorbance at 490 nm, indicating that LDH is only released upon disruption of the cell membrane.

2.13.2 LDH Assay After Exposure to MF and cat-MF

For the assessment of cytotoxicity after cat-MF exposure using the LDH assay, 5000 cells from three different patients were seeded in DMEM medium containing 5% FBS in quadruplicate into wells of a 96 well plate, left to adhere overnight and exposed for 30 min to 125 μL of MF or cat-MF at concentrations ranging from 0.01 to 3 μM, PBS only (untreated control), or a Lysis Buffer provided by the manufacturer as a positive control. The supernatant was removed, cells washed with PBS and cultured in DMEM with reduced serum content (5% FBS) for 24 h after exposure to native MF or cat-MF, because higher serum concentrations are known to interfere with the assay. The assay was performed using the cell culture supernatants as described for the optimisation experiment above.

Percentage cytotoxicity was calculated as follows:

$$\%cytotoxicity = \frac{LDH_{MF} - LDH_{sp}}{LDH_{\max} - LDH_{sp}} * 100 \quad (2.17)$$

where LDH_{MF} was measured in cells exposed to MF/cat-MF, the spontaneous LDH, LDH_{sp}, was measured in the untreated cell control and maximum LDH, LDH_{max}, was measured in cells treated with Lysis Buffer.

For statistical analysis, data were presented as mean ± standard deviation of three biological repeats (total n = 12). Statistical analysis was performed using IBM SPSS Statistics version 21. As for the MTS and Alamar Blue assay, the means of groups treated with MF, cat-MF, and PBS (untreated control) were compared using Two-Way ANOVA, in which effects of dose as well as surface functionalisation (unfunctionalised vs. cationised) were investigated (see Sect. 2.12.2 for rationale behind statistical analysis).

2.14 Long-Term Assessment of Cell Proliferation

Proliferation of hMSCs was assessed over a three week period using a cell counting assay. 150000 cells from three different patients were seeded in culture medium into 75 cm^2 tissue culture flasks and left to adhere overnight. hMSCs were exposed for 30 min to 3 mL of 0.5 μM cat-MF or PBS (untreated cell control). The supernatant was removed, cells washed with PBS and cultured in DMEM for up to three weeks. At weekly intervals, cells were harvested, counted, and population doublings, *PD*, were calculated as follows:

$$PD = \frac{\log\left(\frac{n}{150{,}000}\right)}{\log 2} \tag{2.18}$$

where *n* was the number of counted cells, and 150000 denotes the number of cells initially seeded. Population doublings of the cat-MF treated cells were normalised to the untreated cell control. The population doubling values of cat-MF labelled and untreated hMSCs were compared at each interval using the Wilcoxon Signed-Rank test in IBM SPSS Statistics version 21. A non-parametric test was chosen because of the small sample size (n = 3), which made it difficult to confirm normality for the data sets. The Wilcoxon Signed-Rank test was used because two related samples were compared at each interval (i.e. week one, two and three): the untreated cells derived from three patients *versus* the cat-MF-treated cells derived from the same three patients.

2.15 Adipogenesis of hMSCs Exposed to cat-MF

Theory

Differentiation potential of hMSCs can be assessed by growing cells in a monolayer culture, while supplementing the culture medium with adipogenic growth factors, such as 1-methyl-3-isobutylxanthine, dexamethasone, insulin, and indomethacin [60]. Adipogenesis can be confirmed visually by staining lipid-rich vacuoles within the cells with a fat-soluble dye. Oil Red is a dye that functions as an oil-soluble stain. It is prepared as a solution in isopropanol and partitions into lipid vacuoles upon incubation with adipocytes, because of its increased solubility in fatty substances compared to isopropanol [61].

Method

74000 hMSCs from one patient were seeded in culture medium into the inner wells of a 24 well plate and left to adhere overnight. The outer wells were filled with **antifungal water**, containing 0.1 mg mL^{-1} penicillin and streptomycin, and 2.5 μg mL^{-1} amphotericin B in sterile dH_2O. Cells were exposed to 0.75 mL of 1 μM of cat-MF or PBS (untreated control) for 30 min. The supernatant was removed, cells washed in PBS and cultured in αMEM medium with 10% (v/v) FBS,

1% (v/v) penicillin/streptomycin and 1% (v/v) glutamax solution, containing 10 μl mL^{-1} human adipogenic supplement (StemXVivo, R&D Systems) for three weeks with media changes performed twice a week. After three weeks, adipocyte formation was assessed using Oil Red staining of intracellular lipid vacuoles. Medium was removed, cells were washed with PBS and fixed with 4% (w/v) paraformaldehyde for 30 min at room temperature. The fixative was removed and the cells washed first with PBS followed by 60% (v/v) isopropanol. Oil Red dye was prepared by stirring 5 mg mL^{-1} of Oil Red in isopropanol overnight. The dye was diluted to 3 mg mL^{-1} using dH_2O and then passed through filter paper to remove aggregates. Cells were incubated with 0.5 mL of Oil Red stain for 30 min at room temperature. The stain was aspirated and cells washed with 60% (v/v) isopropanol. Stained monolayers were observed using an inverted light microscope at 10 X magnification (Leica DM IRB).

2.16 Osteogenesis of hMSCs Exposed to cat-MF

Theory

Analogous to adipogenic differentiation, the osteogenic differentiation potential of hMSCs can be assessed by growing cells in a monolayer culture, while supplementing the culture medium with osteogenic growth factors, e.g. dexamethasone, b-glycerol phosphate and ascorbate [60]. Osteogenesis can be confirmed visually by staining calcium deposits. Undifferentiated hMSCs have no extracellular calcium deposits, whereas differentiated osteoblasts produce vast extracellular calcium deposits [60]. Alizarin Red is a dye that forms a complex with calcium ions, resulting in a red product [62].

Method

37000 hMSCs from one patient were seeded in culture medium into the inner wells of a 24 well plate and left to adhere overnight. The remaining outer wells were filled with antifungal water. Cells were exposed to 0.75 mL of 1 μM of cat-MF or PBS (untreated control) for 30 min. The supernatant was removed, cells washed in PBS and cultured in αMEM medium with 10% (v/v) FBS, 1% (v/v) penicillin/streptomycin and 1% (v/v) glutamax solution, containing 50 μg mL^{-1} human osteogenic supplement (StemXVivo, R&D Systems) for three weeks with media changes performed twice a week.

Osteoblast formation was visualised using Alizarin Red staining of calcium phosphate deposits. Cells were washed with PBS and fixed for 1 h at 4 °C using 0.5 mL ice-cold 70% (v/v) ethanol. Alizarin Red dye was prepared by stirring 13.75 mg mL^{-1} of Alizarin Red in dH_2O overnight. The pH was adjusted to 4.1 using potassium hydroxide and the dye was passed through filter paper to remove aggregates. The fixative was removed and the cells were incubated with 0.5 mL alizarin red solution for five minutes at room temperature. The cells were then washed five times with PBS.

2.17 Chondrogenesis of hMSCs Exposed to cat-MF

Theory

Unlike adipogenesis and osteogenesis, chondrogenic differentiation capacity of hMSCs is often investigated in a three dimensional culture, because chondrocytes lose their typical characteristics in monolayer culture [63]. Here, hMSCs were seeded onto fibronectin-coated PGA scaffolds and treated with chondrogenic growth factors such as transforming growth factor–β3 (TGF-β3) [60]. PGA has proven to be one of the most suitable scaffold materials for cartilage growth, [64] and is mostly used in the form of an unwoven fiber mesh (Fig. 2.27).

Method

1000000 hMSCs were seeded in culture medium into 75 cm^2 tissue culture flasks and left to adhere overnight. The cells were exposed to 3 mL of 0.5 μM of cat-MF or PBS for 30 min. The cat-MF or PBS supernatant was removed, cells were washed with PBS and left in culture medium for 24 h. After that, hMSCs were harvested with trypsin/EDTA and counted. 300000 cells were suspended in 30 μL of culture medium and loaded onto fibronectin-coated PGA tissue engineering scaffolds of 5 mm diameter and 2 mm thickness and placed in agarose coated wells of a 24 well plate. Cells were left to attach to the scaffold over night, and then cultured for one week in DMEM containing 4500 mg glucose/L, 1% (v/v) penicillin/streptomycin, 1% (v/v) glutamax, 1% (v/v) sodium pyruvate, and 1% (v/v) insulin-transferrin-sodium selenite, supplemented with 100 nM dexamethasone, 80 nM ascorbic acid and 10 ng/ml TGF-β3. Three media changes were performed during this first week. After that, the medium was additionally supplemented with 10 ng/ml insulin and scaffolds were cultured for a further four weeks

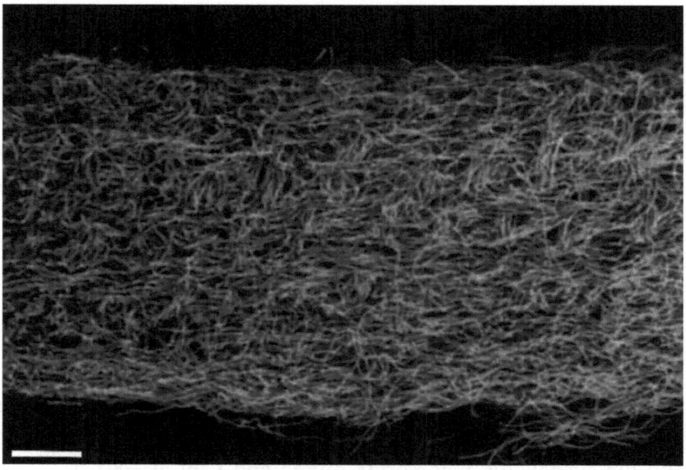

Fig. 2.27 Scanning electron micrograph of a PGA scaffold cross section. *Scale bar* 1 mm. Image re-printed with permission from Freed et al. [65]

with media changes three times a week. After five weeks in total, half of the scaffolds were stored at −80 °C until biochemical analysis, and the other half was fixed in 4% (w/v) paraformaldehyde over night at room temperature, dehydrated in 70% ethanol for 2 h and then submitted to the histology service of the School for Cellular and Molecular Medicine at the University of Bristol. Histological sections of 5 µm thickness were stained with H&E (see Sect. 2.8.2) and Safranin O (see Sect. 2.8.3). They were also stained for iron deposits using Prussian Blue (see Sect. 2.8.4), and for collagen type II using immunostaining (see Sect. 2.8.5). Stained cartilage sections were observed using an inverted light microscope at various magnifications (Leica DM IRB).

The engineered cartilage constructs were also digested for biochemical analysis of type II collagen and glycosaminoglycans (GAG) (see Sects. 2.9.2 and 2.9.3)

2.18 Bacterial Culture

2.18.1 Liquid Culture

T7 Express BL21 carbenicillin-resistant *Escherichia coli* (*E. coli*) bacteria were kindly provided by Dr. James Armstrong (School of Cellular and Molecular Medicine, University of Bristol). 10 mL of sterilised Lysogeny Broth (LB) medium, containing 10 g L^{-1} bacto-tryptone (BD Biosciences, USA), 5 g L^{-1} bacto-yeast extract (BD Biosciences, USA) and 10 g L^{-1} of sodium chloride in dH2O, and supplemented with 10 µL of a 50 µg mL^{-1} carbenicillin solution (Apollo Scientific, UK) was inoculated with a sterile pipette tip dipped into *E. coli* glycerol stocks. This starter culture of bacteria was incubated over night in a rotary shaker at 200 rotations per minute (rpm) and 37 °C and used the next day for experiments. Fresh starter cultures were prepared from glycerol stocks prior to each experiment.

Staphylococcus aureus (*S. aureus*) SH1000 (chloramphenicol resistant) were a kind gift from Ramesh Wigneshweraraj and Andrew Edwards, Imperial College London. Liquid cultures were prepared as described for *E. coli*, but using Nutrient broth (1 g L glucose, 15 g L peptone, 6 g L sodium chloride and 3 g L yeast extract), supplemented with 3.4 µg mL^{-1} chloramphenicol.

2.18.2 Glycerol Stocks

Glycerol stocks were prepared by centrifuging 1 mL of bacterial suspension at 12300 × g for five minutes. Half of the broth supernatant was replaced with glycerol and the suspension was vortex mixed and kept at −80 °C.

2.18.3 Quantification of Bacteria in Water Samples

Theory

In this work, two methods for the quantification of bacteria were explored: measurement of the absorbance of bacterial suspensions using UV/Visible spectroscopy, and spreading samples onto nutrient rich agar plates followed by quantification of the number of colony forming units (cfu) that have grown after 24 h incubation at 37° C.

The plate count method is an indirect measure of cell density and reveals information on the number of live bacteria, whereas measuring the absorbance yields information on the amount of all bacteria in the suspension, dead and alive. For the plate count method a dilution series is prepared such that the bacteria plated out are dilute enough to be counted accurately (between 20 and 200 colonies should be present). It is then assumed that in a dilute bacterial suspension each viable cell is separate from the others and will develop into a discrete colony, such that the number of colonies reflects the number of bacteria that were present in the sample [66]. The *E. coli* strain used in the present experiments is resistant against the antibiotic carbenicillin, and by adding this antibiotic to the nutrient agar it was ensured that only *E. coli* and no other bacteria could grow on the plate.

Increasing absorbance in a bacterial culture is an indicator for bacterial growth. The amount of light scattered by a suspension of bacteria increases as the cell population increases, such that the absorbance indirectly reflects the number of bacteria present (see Sect. 2.2.2). This method is faster than the plate count method, but its sensitivity must be evaluated using a range of bacteria concentrations.

Method

UV/visible spectroscopy of liquid bacterial cultures using absorbance measurements at $\lambda = 600$ nm is described in Sect. 2.2.2.

For the plate counting method, plates were prepared in proximity to the flame of a Bunsen burner to maintain sterile conditions. Approximately 15 mL of molten 1.6% (w/v) LB agar, supplemented with 50 ng mL^{-1} carbenicillin, were poured into sterile plastic petri dishes. The agar was set at 4 °C for at least 1 h. A range of *E. coli* suspensions were prepared and their absorbance measured as described in Sect. 2.2.2. These suspensions were further diluted in a series of 1 in 10, and 0.1 mL of the lowest three concentrations were plated out in triplicate using sterile, L-shaped plastic spreaders (VWR, UK). The plates were incubated overnight at 37 °C and cfu counted on the plates that contained 20 to 200 colonies. Bacterial concentration in cfu mL^{-1} was determined by taking the dilution factor into consideration.

E. coli concentration could be related to a measurable absorbance values down to approximately 10^7 cfu mL^{-1} (Fig. 2.28). Below this value absorbance readings were not sensitive enough to accurately determine the number of *E. coli* present, because of insufficient scattering from the dilute cell suspension. Therefore, the plating method was chosen to quantify *E. coli* in the subsequent decontamination experiments. However, absorbance measurements were used to adjust bacterial

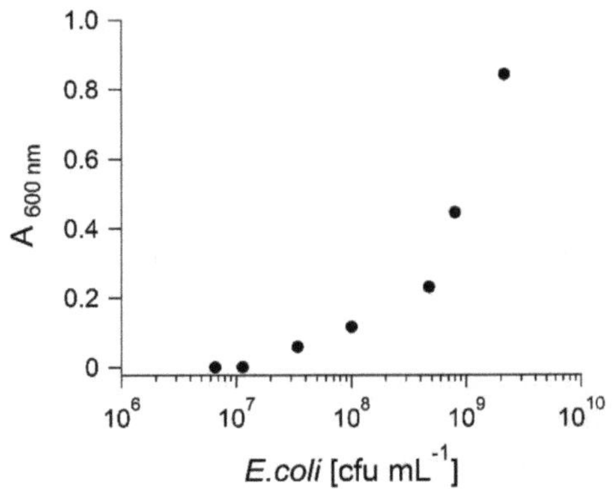

Fig. 2.28 Absorbance measured at 600 nm as a function of *E. coli* present in the water sample as determined by the plate count method. Below 10^7 cfu mL^{-1} absorbance readings are not sensitive enough to accurately determine cell number

concentration in inoculated water samples prior to magnetic separation to approximately 10^8 cfu mL^{-1}, which corresponded to an absorbance value of around 0.3. Assessing absorbance of a bacterial suspension was first introduced in 1907 [67] and has since widely been used to achieve suspensions with a specific concentration of bacteria [68].

2.19 Magnetic Labelling of *E. coli*

All magnetisation experiments were conducted in proximity to the flame of a Bunsen burner to keep sterile conditions. The plate counting method (see Sect. 2.18.3) was used throughout to determine bacterial concentration before and after MACS.

2.19.1 Rapidity and Extent of E. coli *Magnetisation*

A 1 mL sample of liquid starter culture was centrifuged for 5 min at 14000×g, the medium supernatant removed, and the pellet re-suspended in 1 mL of sterile dH_2O. This suspension was diluted with sterile dH_2O to achieve a bacterial suspension with an absorbance of approximately 0.3. 0.3 mL of this suspension was mixed with an equal volume of sterile dH_2O or 1 μM of cat-MF (sterile filtered) to yield a final cat-MF concentration of 0.5 μM. The mixture was agitated briefly, and then incubated at room temperature for 1, 5, 15 and 30 min. After each incubation period, 0.5 mL of the bacterial suspension was loaded onto MACS columns and the flow-through was collected.

To **determine cfu mL^{-1}** in the bacterial suspension before MACS, samples of the initial bacterial suspension were diluted $1/10^5$, $1/10^6$ and $1/10^7$. 0.1 mL samples of these dilutions were plated out in triplicate and incubated at 37 °C overnight. To determine the number of cfu in the water after MACS, samples of the flow-through were diluted 1/10 to $1/10^2$, plated out in triplicate and incubated at 37 °C overnight. After incubation, the number of colonies on the agar plates was counted and cfu mL^{-1} determined by taking into account the respective dilution factors.

The **percentage removal** was calculated as follows:

$$removal(\%) = 100 - \left(\frac{c_{after}}{c_{before}} * 100\right) \tag{2.19}$$

where c_{after} was the bacterial concentration in the flow-through after MACS, and c_{before} was the bacterial concentration in the initial bacterial suspension before MACS.

2.19.2 Magnetisation of E. coli *with Different cat-MF Concentrations*

A 1 mL sample of liquid starter culture was centrifuged for 5 min at 14000×g, the medium supernatant removed, and the pellet re-suspended in 1 mL of sterile dH_2O. This suspension was diluted in sterile dH_2O to achieve a bacterial suspension with an absorbance of approximately 0.3. 0.3 mL of this suspension was mixed with an equal volume of sterile water or various concentrations of cat-MF (sterile filtered) to yield a final cat-MF concentration of 0.01 to 1 μM cat-MF. The mixture was agitated briefly, and then incubated at room temperature for 15 min. After the incubation period, 0.5 mL of the bacterial suspension was loaded onto MACS columns, the flow-through was collected and cfu mL^{-1} and percentage removal determined as described in Sect. 2.19.1.

2.19.3 Magnetic Capture of Various Concentrations of E. coli

A 1 mL sample of liquid starter culture was centrifuged for 5 min at 14000 × g, the medium supernatant removed, and the pellet re-suspended in 1 mL of sterile dH_2O. The suspension was diluted in sterile dH_2O to achieve a bacterial suspension with an absorbance of approximately 0.3. This suspension was further diluted in a series of 1 in 10, down to a dilution of $1/10^6$. 0.3 mL of each dilution was incubated with an equal volume of 1 μM of cat-MF (sterile filtered) to yield a final cat-MF concentration of 0.5 μM. The mixture was agitated briefly, and then incubated at room temperature for 15 min. 0.5 mL of the bacterial suspension were loaded onto MACS columns, the flow-through was collected and cfu mL^{-1} and percentage removal determined as described in Sect. 2.19.1.

2.19.4 Concentration of Low Amounts of E. coli

A 1 mL sample of liquid starter culture was centrifuged for 5 min at 14,000×g, the medium supernatant removed, and the pellet re-suspended in 1 mL of sterile dH_2O. This suspension was diluted in sterile dH_2O to achieve a bacterial suspension with an absorbance of approximately 0.3. This suspension was further diluted in a series of 1 in 10, down to a dilution of $1/10^7$. 0.3 mL of this dilution was incubated with an equal volume of 1 μM of cat-MF (sterile filtered) to yield a final cat-MF concentration of 0.5 μM. The mixture was agitated briefly, and then incubated at room temperature for 15 min. 2 mL of the bacterial suspension were loaded onto MACS columns, the flow-through was collected. The column was removed from the magnet and any bacteria immobilised in the column were eluted with 0.4 mL of sterile dH_2O using a plunger. The initial solution, the flow through and the eluted bacterial suspension were plated out without further dilution. cfu mL^{-1} and percentage removal were determined as described in Sect. 2.19.1.

2.20 Comparing Magnetic Capture of *E. coli* and *S. aureus*

1 mL samples of liquid starter culture of *E. coli* and *S. aureus* were centrifuged for 5 min at 14000 × g, the medium supernatant removed, and the pellets re-suspended in 1 mL of sterile dH_2O. The suspensions were diluted in sterile dH_2O to achieve a bacterial suspension with an absorbance of approximately 0.3. 0.3 mL of each suspension was mixed with an equal volume of sterile dH_2O or 1 μM of cat-MF (sterile filtered) to yield a final cat-MF concentration of 0.5 μM. The mixtures were agitated briefly, and then incubated at room temperature for 15 min. 0.5 mL of each bacterial suspension were loaded onto MACS columns, the flow-through collected and cfu mL^{-1} and percentage removal determined as described in Sect. 2.19.1. Note that *S. aureus* were plated out on Nutrient agar rather than LB agar.

2.21 Assessing *E. coli* Growth in the Presence of cat-MF

A 0.5 mL sample of liquid starter culture was diluted in LB broth medium containing 50 ng mL^{-1} carbenicillin to achieve a bacterial suspension with an absorbance of approximately 0.1. 10 μL of this *E. coli* suspension was added to 180 μL of carbenicillin supplemented LB broth containing 0.01 to 1 μM cat-MF. The absorbance of the bacterial solution, as well as of the background solutions, was recorded every 10 min at over 4 h at 37 °C. The absorbance of the background solutions were subtracted from the bacterial suspensions to account for any absorbance effects due to the presence of cat-MF particles.

References

1. Scorpio, R.: Fundamentals of Acids, Bases, Buffers and Their Application to Biochemical Systems. Kendall/Hunt Publishing Company (2000)
2. Förster, H.: Uv/Vis Spectroscopy. In Characterization I, pp 337–426. Springer (2004)
3. Koch, A.L.: Turbidity measurements of bacterial cultures in some available commercial instruments. Anal. Biochem. **38**(1), 252–259 (1970)
4. Goodwin, T., Morton, R.: The spectrophotometric determination of tyrosine and tryptophan in proteins. Biochem. J. **40**(5–6), 628 (1946)
5. Lakowicz, J.R.: Principles of Fluorescence Spectroscopy. Springer Science & Business Media (2013)
6. Bradford, M.M.: A rapid and sensitive method for the quantitation of microgram quantities of protein utilizing the principle of protein-dye binding. Anal. Biochem. **72**(1), 248–254 (1976)
7. Sober, H.A., Gutter, F.J., Wyckoff, M.M., Peterson, E.A.: Chromatography of proteins. Ii. Fractionation of serum protein on anion-exchange cellulose. J. Am. Chem. Soc. **78**(4), 756–763 (1956)
8. Wu, C.-S.: Handbook of Size Exclusion Chromatography and Related Techniques: Revised and Expanded, vol. 91. CRC Press (2003)
9. Boyes, E., Gai, P.: Environmental high resolution electron microscopy and applications to chemical science. Ultramicroscopy **67**(1), 219–232 (1997)
10. Brakenhoff, G., Blom, P., Barends, P.: Confocal scanning light microscopy with high aperture immersion lenses. J. Microscopy (Oxford) **117**(2), 219–232 (1979)
11. Brenner, S., Horne, R.: A negative staining method for high resolution electron microscopy of viruses. Biochem. Biophys. Acta. **34**, 103–110 (1959)
12. Hopwood, D.: Theoretical and practical aspects of glutaraldehyde fixation. Histochem. J. **4**(4), 267–303 (1972)
13. Bancroft, J.D., Gamble, M.: Theory and Practice of Histological Techniques. Elsevier Health Sciences (2008)
14. Bernhard, W.: A new staining procedure for electron microscopical cytology. J. Ultrastruct. Res. **27**(3), 250–265 (1969)
15. Daddow, L.: An abbreviated method of the double lead stain technique. J. Submicrosc. Cytol. **18**(1), 221–224 (1986)
16. Pecora, R.: Dynamic light scattering measurement of nanometer particles in liquids. J. Nanopart. Res. **2**(2), 123–131 (2000)
17. Goldburg, W.: Dynamic light scattering. Am. J. Phys. **67**(12), 1152–1160 (1999)
18. O'Brien, R.W., White, L.R.: Electrophoretic mobility of a spherical colloidal particle. J. Chem. Soc. Faraday Trans. 2: Mol. Chem. Phys. **74**, 1607–1626 (1978)
19. Miller, J.F., Schätzel, K., Vincent, B.: The determination of very small electrophoretic mobilities in polar and nonpolar colloidal dispersions using phase analysis light scattering. J. Colloid Interface Sci. **143**(2), 532–554 (1991)
20. Hillenkamp, F., Karas, M., Beavis, R.C., Chait, B.T.: Matrix-assisted laser desorption/ionization mass spectrometry of biopolymers. Anal. Chem. **63**(24), 1193A–1203A (1991)
21. Opsal, R.B., Owens, K.G., Reilly, J.P.: Resolution in the linear time-of-flight mass spectrometer. Anal. Chem. **57**(9), 1884–1889 (1985)

22. Hou, X., Jones, B.T.: Inductively coupled plasma-optical emission spectrometry. Encyclopedia of Analytical Chemistry (2000)
23. Mercereau, J.: Superconducting magnetometers. Revue de physique appliquée **5**(1), 13–20 (1970)
24. Clarke, J.: Principles and applications of squids. Proc. IEEE **77**(8), 1208–1223 (1989)
25. Pykett, I.L., et al.: Principles of nuclear magnetic resonance imaging. Radiology **143**(1), 157–168 (1982)
26. Bloembergen, N., Purcell, E.M., Pound, R.V.: Relaxation effects in nuclear magnetic resonance absorption. Phys. Rev. **73**(7), 679 (1948)
27. Rehr, R.B., et al.: Improved in vivo magnetic resonance imaging of acute myocardial infarction after intravenous paramagnetic contrast agent administration. Am J. Cardiol. **57** (10), 864–868 (1986)
28. Bydder, G., Hajnal, J., Young, I.: MRI: use of the inversion recovery pulse sequence. Clin. Radiol. **53**(3), 159–176 (1998)
29. Okuda, M., Eloi, J.C., Sarua, A., Jones, S.E.W., Schwarzacher, W.: Energy barrier distribution for dispersed mixed oxide magnetic nanoparticles. J. Appl. Phys. **111**(7) (2012)
30. Okuda, M., Eloi, J.-C., Jones, S.E.W., Sarua, A., Richardson, R.M., Schwarzacher, W.: Fe3o4 Nanoparticles: protein-mediated crystalline magnetic superstructures. Nanotechnology **23** (41), 415601 (2012)
31. Danon, D., Skutelsk, E., Marikovs, Y., Goldstei, L.: Use of cationized ferritin as a label of negative charges on cell surfaces. J. Ultrastruct. Res. **38**(5–6), 500–510 (1972)
32. Perriman, A.W., Cölfen, H., Hughes, R.W., Barrie, C.L., Mann, S.: Solvent-free protein liquids and liquid crystals. Angew. Chem. Int. Ed. **48**(34), 6242–6246 (2009)
33. Fox, C.H., Johnson, F.B., Whiting, J., Roller, P.P.: Formaldehyde fixation. J. Histochem. Cytochem. **33**(8), 845–853 (1985)
34. Fischer, A.H., Jacobson, K.A., Rose, J., Zeller, R.: Hematoxylin and eosin staining of tissue and cell sections. Cold Spring Harbor Protocols (5), pdb. prot4986 (2008)
35. Marshall, P., Horobin, R.: The mechanism of action of "Mordant" dyes—a study using preformed metal complexes. Histochemie **35**(4), 361–371 (1973)
36. Waheed, A., Rao, K.S., Gupta, P.: Mechanism of dye binding in the protein assay using eosin dyes. Anal. Biochem. **287**(1), 73–79 (2000)
37. Rosenberg, L.: Chemical basis for the histological use of safranin o in the study of articular cartilage. J. Bone Joint Surg. **53**(1), 69–82 (1971)
38. Martin, I., Obradovic, B., Freed, L.E., Vunjak-Novakovic, G.: Method for quantitative analysis of glycosaminoglycan distribution in cultured natural and engineered cartilage. Ann. Biomed. Eng. **27**(5), 656–662 (1999)
39. Kiernan, J.A.: Anionic counterstains. Cold Spring Harbor Protocols (7), pdb. top51 (2008)
40. Bruns, R.R., Palade, G.E.: Studies on blood capillaries Ii. Transport of ferritin molecules across the wall of muscle capillaries. J. Cell Biol. **37**(2), 277–299 (1968)
41. Sundberg, D.R., Broman, H.: The application of the prussian blue stain to previously stained films of blood and bone marrow. Blood **10**(2), 160–166 (1955)
42. Schöpf, B., et al.: Methodology description for detection of cellular uptake of PVA coated superparamagnetic iron oxide nanoparticles (Spion) in synovial cells of sheep. J. Magn. Magn. Mater. **293**(1), 411–418 (2005)
43. Frank, M., Dapson, R., Wickersham, T., Kiernan, J.: Certification procedures for nuclear fast red (Kernechtrot), Ci 60760. Biotech. Histochem. **82**(1), 35–39 (2007)
44. Sternberger, L.A., Hardy, P.H., Cuculis, J.J., Meyer, H.G.: The unlabeled antibody enzyme method of immunohistochemistry preparation and properties of soluble antigen-antibody complex (Horseradish Peroxidase-Antihorseradish Peroxidase) and its use in identification of spirochetes. J. Histochem. Cytochem. **18**(5), 315–333 (1970)
45. Seligman, A.M., Karnovsky, M.J., Wasserkrug, H.L., Hanker, J.S.: Nondroplet ultrastructural demonstration of cytochrome oxidase activity with a polymerizing osmiophilic reagent, diaminobenzidine (Dab). J. Cell Biol. **38**(1), 1–14 (1968)

46. Olsen, J.V., Ong, S.-E., Mann, M.: Trypsin cleaves exclusively c-terminal to arginine and lysine residues. Mol. Cell. Proteomics **3**(6), 608–614 (2004)
47. Kurachi, K., Powers, J.C., Wilcox, P.E.: Kinetics of the reaction of chymotrypsin Aα with peptide chloromethyl ketones in relation to its subsite specificity. Biochemistry **12**(4), 771–777 (1973)
48. Smythe, C.: The reaction of iodoacetate and of iodoacetamide with various sulfhydryl groups, with urease, and with yeast preparations. J. Biol. Chem. **114**(3), 601–612 (1936)
49. Umezawa, H., AOYAGI, T., Morishima, H., Matsuzaki, M., Hamada, M., Takeuchi, T.: Pepstatin, a New Pepsin Inhibitor Produced by Agtinomygetes. J. Antibiotics **23**(5), 259–262 (1970)
50. Diermayr, P., Kroll, S., Klostermeyer, H.: Influence of Edta and metal ions on a metalloproteinase from pseudomonas fluorescens biotype I. Bio. Chem. Hoppe-Seyler **368** (1), 57–62 (1987)
51. Eyre, D.: Collagen of articular cartilage. Arthrit. Res. **4**(1), 30–35 (2002)
52. Hollander, A.P., et al.: Increased damage to type ii collagen in osteoarthritic articular cartilage detected by a new immunoassay. J. Clin. Investig. **93**(4), 1722 (1994)
53. Roughley, P.J., Lee, E.R.: Cartilage proteoglycans: structure and potential functions. Microsc. Res. Tech. **28**(5), 385–397 (1994)
54. Müller, G., Hanschke, M.: Quantitative and qualitative analyses of proteoglycans in cartilage extracts by precipitation with 1, 9-Dimethylmethylene blue. Connect. Tissue Res. **33**(4), 243–248 (1996)
55. Stone, J.E., Akhtar, N., Botchway, S., Pennock, C.A.: Interaction of 1, 9-Dimethylmethylene blue with glycosaminoglycans. Ann. Clin. Biochem.: Int. J. Biochem. Med. **31**(2), 147–152 (1994)
56. Baeuerle, P.A., Huttner, W.B.: Chlorate–a potent inhibitor of protein sulfation in intact cells. Biochem. Biophys. Res. Commun. **141**(2), 870–877 (1986)
57. Berridge, M.V., Tan, A.S.: Characterization of the cellular reduction of 3-(4, 5-Dimethylthiazol-2-Yl)-2, 5-Diphenyltetrazolium Bromide (Mtt): subcellular localization, substrate dependence, and involvement of mitochondrial electron transport in mtt reduction. Arch. Biochem. Biophys. **303**(2), 474–482 (1993)
58. Candeias, L., MacFarlane, D.S., McWhinnie, S.W., Maidwell, N., Roeschlaub, C., Sammes, P.: The catalysed nadh reduction of resazurin to resorufin. J. Chem. Soc. Perkin Trans. **2**(11), 2333–2334 (1998)
59. Decker, T., Lohmann-Matthes, M.-L.: A quick and simple method for the quantitation of lactate dehydrogenase release in measurements of cellular cytotoxicity and tumor necrosis factor (TNF) activity. J. Immunol. Methods **115**(1), 61–69 (1988)
60. Pittenger, M.F., et al.: Multilineage potential of adult human mesenchymal stem cells. Science **284**(5411), 143–147 (1999)
61. Ramirez-Zacarias, J., Castro-Munozledo, F., Kuri-Harcuch, W.: Quantitation of adipose conversion and triglycerides by staining intracytoplasmic lipids with oil red O. Histochemistry **97**(6), 493–497 (1992)
62. Puchtler, H., Meloan, S.N., Terry, M.S.: On the history and mechanism of alizarin and Alizarin Red S stains for calcium. J. Histochem. Cytochem. **17**(2), 110–124 (1969)
63. Aulthouse, A.L., et al.: Expression of the human chondrocyte phenotype in vitro. In Vitro Cell. Dev. Biol. **25**(7), 659–668 (1989)
64. Grande, D., Halberstadt, C., Naughton, G., Schwartz, R., Manji, R.: Evaluation of matrix scaffolds for tissue engineering of articular cartilage grafts. J. Biomed. Mater. Res. **34**(2), 211–220 (1997)
65. Freed, L.E., et al.: Biodegradable polymer scaffolds for tissue engineering. Biotechnol. (N Y). **12**(7), 689–693 (1994)
66. Zuberer, D.A.: Recovery and enumeration of viable bacteria. Methods of Soil Analysis: Part 2—Microbiological and Biochemical Properties (methodsofsoilan2), pp. 119–144 (1994)

67. McFarland, J.: The nephelometer: an instrument for estimating the number of bacteria in suspensions used for calculating the opsonic index and for vaccines. J. Am. Med. Assoc. **49** (14), 1176–1178 (1907)
68. Sutton, S.: Measurement of microbial cells by optical density. J. Valid. Technol. **17**, 47–49 (2011)

Chapter 3
Characterising Magnetoferritin and Cationised Magnetoferritin

In this chapter, the physico-chemical characteristics of MF and cat-MF are presented. Cationisation of MF has not yet been explored as a functionalisation route, but cationisation of ferritin has first been described in the early 1970s [1]. The results presented here have been obtained using a method developed from previously reported ferritin cationisation protocols [1, 2].

To determine particle size, TEM and DLS were employed. The composition of the MF core was analysed using ICP-OES, and cationisation efficiency was analysed using zeta potentiometry and MALDI-TOF mass spectrometry. Finally, the magnetic properties of both MF and cat-MF were measured using SQUID magnetometry and MRI.

3.1 Nanoparticle Size and Composition

To confirm nanoparticle mineralisation inside the ApoF cavity and determine the size of this core, TEM is a suitable technique because metal-based nanoparticles appear as dark, electron-dense structures in the image. However, to visualise the protein shell, staining is required. Here, 2% phosphotungstic acid was used as a negative stain that surrounds the protein such that it appears as a bright structure on a dark background, as has been shown in previous studies on MF mineralisation. [3–6] Thus it was possible to confirm the presence of nanoparticles within the protein cage (Fig. 3.1). Image analysis of unstained MF samples gave an average core diameter of 5.3 ± 1.1 nm (Fig. 3.1a, d), which is slightly smaller compared to previously reported diameters of cobalt-doped iron oxide MF cores ($\sim$7 nm) [3, 5]. Stained samples gave an average diameter of 12.6 ± 1.1 nm for the entire MF nanoparticle (core and shell) (Fig. 3.1b, e), and 12.7 ± 1.2 nm for cat-MF (Fig. 3.1c, f).

S. Correia Carreira, *Rapid Cell Magnetisation Using Cationised Magnetoferritin*,
Springer Theses, DOI 10.1007/978-3-319-60333-9_3

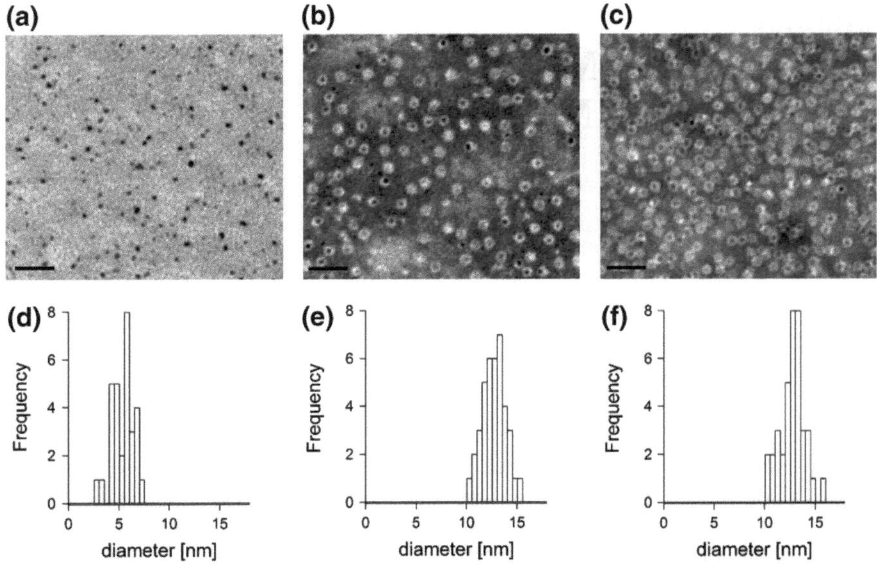

Fig. 3.1 TEM analysis of MF and cat-MF. TEM images of **a** unstained MF, and **b** MF and **c** cat-MF stained with 2% phosphotungstic acid. Mineralised nanoparticle cores are visible as *black dots*, while the protein shell appears as a light ring around the core on the stained images. **d** Image analysis of nanoparticle cores, and **e–f** cores with surrounding protein shell. Scale bars: 50 nm

Detailed analysis of the structure of the MF core have not been undertaken here, but previous studies using selected-area electron diffraction images indicate the presence of both the inverse spinel structure of magnetite (Fe_3O_4) or maghemite (γ-Fe_2O_3) and the spinel structure of Co_3O_4 in the MF core [3]. Furthermore, Raman spectra revealed peaks attributed to $CoFe_2O_4$, Fe_3O_4 and small amounts of γ-Fe_2O, indicating that the core consists of a cobalt ferrite ($Co_xFe_{3-x}O_4$) with 15% maghemite [5].

If TEM is used for size characterisation of nanoparticles which are to be used in suspension, it is also important to complement this technique with others, such as DLS. TEM can only be performed on dried samples, and the drying process may introduce artefacts. DLS, however, determines the hydrodynamic diameter of particles in situ, i.e. suspended in a liquid. The measurement is based on assessing the Brownian motion of the particles (see Sect. 2.4.2). Particle motion is captured in form of a correlation decay function, in which the decay rate is faster for smaller, faster moving particles, compared to larger, slower moving ones. Plotting the correlation function of MF and cat-MF reveals that the correlation decay is faster for MF compared to cat-MF (Fig. 3.2a), indicating an increase in hydrodynamic radius of the protein after cationisation. This is likely due to additional steric bulk of coupled DMPA, which may impair particle mobility. From this correlation function, the instrument software has calculated a diffusion coefficient, z-average and

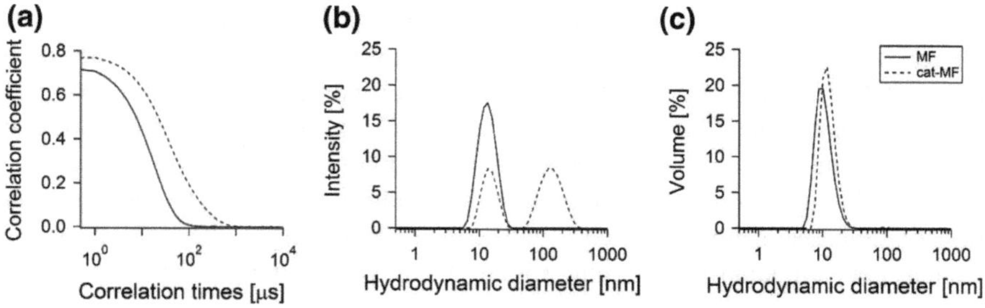

Fig. 3.2 DLS analysis of MF and cat-MF. From the **a** correlation function, decay rates can be extracted for a number of size classes and a size distribution can be generated. This can be displayed as **b** a plot of the relative intensity of light scattered by particles against various size classes, which in turn can be transformed to **c** a volume distribution that takes into account the relative amounts of particles in each intensity peak (see also Table 3.1)

Table 3.1 Size parameters of MF and cat-MF calculated from DLS data

					Volume distribution data			
	D[a] (μm^2/s)	Z[b] (nm)	PDI[c]	Peak 1 diameter (nm)	Peak 1 volume (%)	Peak 2 diameter (nm)	Peak 2 volume (%)
MF	37.2	12.6	0.1	11.8 ± 1.1	100	0	0
cat-MF	14.7	33.4	0.5	12.5 ± 1.4	99.5	118.6 ± 56.0	0.5

[a]Diffusion coefficient
[b]Z-average
[c]Polydispersity index

polydispersity index for the samples, which are summarised in Table 3.1. The z-average is an intensity-weighted mean that is specific to light scattering. It is very sensitive to large aggregates, because the intensity of the scattered light is proportional to the sixth power of the particle diameter. Indeed, the z-average of cat-MF is more than twice as large as that of MF (Table 3.1), which can be explained by the presence of large aggregates of around 100 nm in the cat-MF sample evident in the intensity distribution plot (Fig. 3.2b). The DLS instrument software enables transformation of the intensity distribution to a volume distribution, which reveals that the relative amount of large particle aggregates in the second intensity peak makes up a very small fraction of the whole cat-MF sample (Table 3.1), so that most of the sample volume is in fact occupied by particles with a hydrodynamic diameter of around 12.5 nm (Fig. 3.2c).The polydispersity index (PDI) is a dimensionless figure, ranging from zero to one, where values close to zero indicate a narrow size distribution, and values close to 1 a very polydisperse sample. The PDI is smaller for MF compared to cat-MF, indicating that cationisation leads to an increase in polydispersity of the sample. This could be due to inhomogeneous cationisation, with some proteins being more heavily cationised than others.

To determine the amount of iron and cobalt in the MF cores, inductively-coupled plasma optical emission spectrometry (ICP-OES) was performed on samples of 10 different MF batches. The measured average iron content was 2.65 ± 0.62 ppm and the cobalt content was 0.03 ± 0.01 ppm. Taking into account the amount of protein in the samples, this corresponds to a mass of 45 ± 8 µg of iron and 0.43 ± 0.17 µg of cobalt per milligram of MF, indicating a 1% (w/w) doping of the iron oxide-based nanoparticle core with cobalt. Accordingly, the iron and cobalt content per individual MF protein unit are 3.4×10^{-8} and 3.2×10^{-10} pg.

3.2 Cationisation of MF

To assess the cationisation efficiency of covalent DMPA-coupling to the MF surface, zeta potentiometry was used to measure the surface potential of cat-MF compared to MF, and MALDI-TOF mass spectrometry was used to determine the number of DMPA molecules conjugated to each protein subunit.

The zeta potential changed from −10.5 mV for MF to 8.3 mV for cat-MF after cationisation, confirming a change in surface potential from negative to positive (Fig. 3.3a). Time course cationisation studies using ferritin showed that maximum zeta potential was reached after 2 h crosslinking time (Fig. 3.3b), and that further incubation did not result in additional cationisation. This could be due to the instability of the active ester intermediate formed during the first step of the conjugation reaction, which leads to its hydrolysis and restoration of the carboxyl group (see Sect. 1.9).

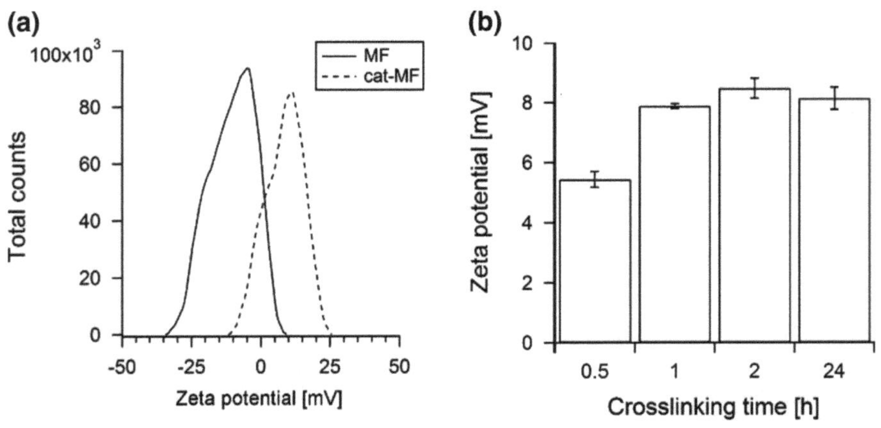

Fig. 3.3 Zeta potentiometry of MF and cat-MF. **a** Zeta potential distribution of MF and cat-MF. **b** cationisation time course study revealing that maximum zeta potential is reached 2 h after addition of the EDC crosslinker. Mean and standard deviation of three zeta potential measurements are shown

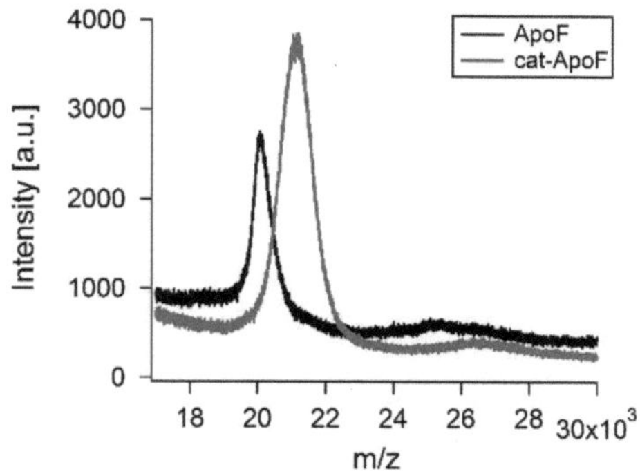

Fig. 3.4 Mass spectrometry analysis of apoferritin and cationised apoferritin subunits. Mass-to-charge spectrum of apoferritin (ApoF) and cationised apoferritin (cat-ApoF) generated by MALDI-TOF. **a** mass increase from 20.1–21.1 kDa is observed after cationisation

Mass spectrometry experiments found a single peak at m/z 20.1 for ApoF, which lies between the literature value for the light and the heavy subunit (19 and 21 kDa, respectively) [7]. For cat-ApoF a subunit weight of 21.1 kDa was found, which corresponds to a mass increase of 1000 Da per subunit (Fig. 3.4).

The molecular mass of a single DMPA molecule is 102 Da, and during the peptide bond formation between the protein carboxyl group and the amine group of DMPA, an oxygen and two hydrogen atoms are shed, resulting in a total mass loss of 18 Da per DMPA molecule coupled. Therefore, for every conjugated DMPA molecule a mass increase of 84 Da is expected. The increase in subunit weight after cationisation was 1000 Da, which corresponds to approximately 12 coupled DMPA molecules per protein subunit, and the cationisation of 288 residues on the entire 24-subunit protein. This is consistent with previous cationisation efficiencies obtained using horse spleen ferritin [2]. The number of cationisable sites, i.e. glutamate and aspartate residues, on the light ferritin chain is 27 (see Sect. 1.9.1), so there is potential to increase the cationisation efficiency. As mentioned earlier, the activated ester intermediate formed during the EDC crosslinking reaction is very unstable and hydrolyses within minutes, thus, not all available DMPA molecules may be coupled to all available carboxyl groups before the inactivation occurs. Therefore, multiple additions of EDC at specific time intervals may help increase cationisation efficiency. If this route is chosen, DMPA content should be maintained at stoichiometric excess to avoid crosslinking between protein units.

3.3 Magnetic Properties

Magnetic saturation and susceptibility of MF and cat-MF were determined by measuring the sample's magnetic moment in response to a field sweep over 2 T at 300 K, which generated a magnetisation curve (Fig. 3.5a). The magnetic moment of the sample increased with increasing external magnetic field strength as a result of the alignment of the sample's magnetic spins with the applied field. When all spins were aligned, the magnetic moment did not increase further, i.e. it reached

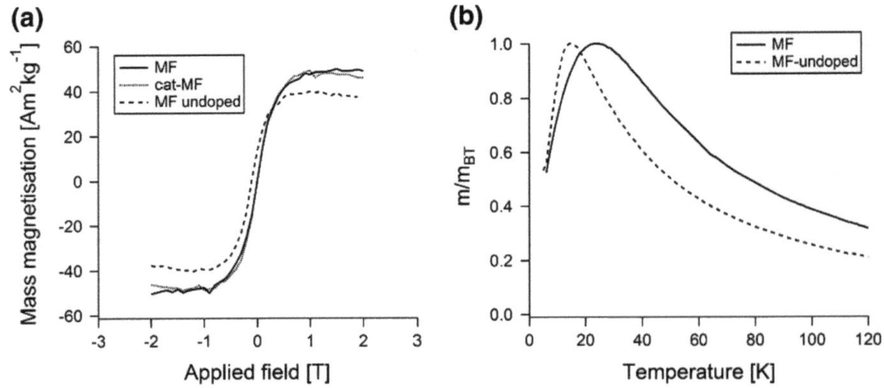

Fig. 3.5 Magnetic characterisation using SQUID magnetometry. **a** Magnetisation curve of MF, cat-MF and un-doped MF (iron oxide core only) at 300 K. The saturation moment was determined by fitting the Langevin function, and mass susceptibility was calculated from the linear portion of the magnetisation curve at fields up to ± 0.3 T. **b** Magnetic moment normalised for magnetic moment at blocking temperature (m_{BT}) of MF and un-doped MF as a function of temperature

saturation. The shape of the curve in response to the applied magnetic field was characteristic for superparamagnetic nanoparticles in that no hysteresis, and thus no remanence or coercivity, were discernible above the blocking temperature [8, 9]. Magnetic saturation and susceptibility were similar for MF and cat-MF, indicating that cationisation had negligible impact on the magnetic properties of the enclosed SPION (Fig. 3.5 and Table 3.2). The blocking temperature was determined by cooling the sample down from 120 to 6 K without applying a magnetic field. This meant that to start with, the magnetic moments of the nanoparticle cores were randomly aligned in the sample. Then, a small magnetic field was applied (0.01 T) and the magnetic moment was recorded upon warming the sample. As the temperature increased, the thermal energy started to disturb the system such that more and more magnetic moments acquired the energy to align with the applied field. The number of aligned moments reached a maximum at the blocking temperature. Above the blocking temperature, the thermal energy became strong enough to randomise the magnetic moments again, which led to a decrease in magnetisation (Fig. 3.5b).

The saturation moment was higher compared to values previously reported for iron oxide nanoparticles of similar size [10, 11]. However, MF containing an iron oxide core only (no cobalt doping) was found to have a saturation moment in the reported range of 40–45 Am^2/kg. Not only did cobalt-doping increase the saturation moment, but also the magnetic susceptibility and the blocking temperature (Table 3.2). An increase in blocking temperature due to cobalt-doping of MF has previously been reported [3, 5], and has been attributed to enhanced magneto crystalline anisotropy of the magnetite core due to cobalt atoms being incorporated in the Fe_3O_4 lattice [12]. An increase in magnetocrystalline anisotropy results in an increased energy barrier for magnetic reversal, explaining the higher blocking

Table 3.2 Magnetic properties of MF and cat-MF and MF without cobalt doping (iron oxide only)

	MF	Cat-MF	Un-doped MF
Magnetic saturation moment (Am^2/kg)	54.9 ± 1.6	55.3 ± 1.4	44.6 ± 1.0
Mass susceptibility ($\times 10^{-4}$ m^3/kg)	1.75 ± 0.08	1.75 ± 0.07	1.4 ± 0.05
Blocking temperature (K)	30	n.d.	15
Longitudinal relaxivity r1 (mM^{-1} s^{-1})	2.6 ± 0.1	2.3 ± 0.1	n.d.
Transverse relaxivity r2 (mM^{-1} s^{-1})	44.6 ± 1.0	52.8 ± 0.8	n.d.

n.d. not determined

temperature of cobalt-doped MF. An interesting feature of the curves in Fig. 3.5b is the broader peak observed for the cobalt-doped MF compared to the un-doped MF. Seeing that a distribution of particle sizes leads to a distribution of blocking temperatures (because blocking temperature depends on particle volume, see Sect. 1.1.5), it follows that introduction of cobalt leads to the mineralisation of less uniformly sized nanoparticle cores.

Longitudinal and transverse relaxivity (r1 and r2, respectively) were determined using a series of spin echo sequences in a 3 T MRI scanner (Fig. 3.6). Relaxivity is defined as the relaxation rate ($T1^{-1}$ or $T2^{-1}$) of nuclear proton spins per unit concentration of contrast agent, and thus it reflects the effect contrast agents have on the relaxation of surrounding proton spins. As mentioned previously (see Sect. 2.4.7), two relaxation mechanisms exist, spin-spin relaxation and spin-lattice relaxation. Both mechanisms have characteristic time constants: spin-spin relaxation occurs within a few hundred milliseconds and is described by T2, whereas spin-lattice relaxation takes several seconds and is described by T1 [13]. When contrast agents are applied, they can influence these time constants locally, thereby altering the image contrast in particular areas. Size, shape, composition, surface properties and degrees of aggregation can influence a contrast agent's relaxivity values. [14–16] For MF and cat-MF, r2 was much higher than r1 (Table 3.2), and the ratio of r2 to r1 was 17 for MF and 23 for cat-MF. This indicates that MF and cat-MF have good contrast enhancing properties in T2- weighted images, in which they cause significant local signal loss (darkening of the image). This is expected for spherical SPION-based contrast agents, which affect spin-spin relaxation to a higher degree than spin-lattice relaxation [17, 18]. Spin-spin relaxation is driven by local magnetic field inhomogeneities caused by neighbouring proton spins, or in fact by the presence of SPIONs that become highly magnetised in the applied magnetic field of the MRI scanner and create significant inhomogeneities.

Comparing r1 and r2 of MF and cat-MF to other SPIONs of similar size, both values were found to be lower than commercially available ultrasmall SPIONs that have reported r1 and r2 values of 6 and 120 $mM^{-1}s^{-1}$, respectively (at 3 T) [19]. However, r2 was found to be approximately 45% higher than previously reported

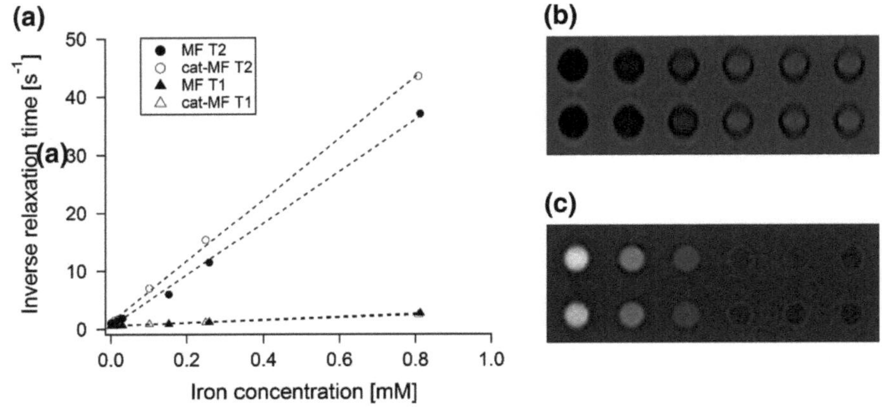

Fig. 3.6 Relaxivity of MF and cat-MF. **a** Inverse longitudinal (T1) and transverse (T2) relaxation times of MF and cat-MF plotted as a function of iron concentration. The slope of the linear fit represents the relaxivity (r) of the superparamagnetic core. **b** Representative spin echo image of a concentration range of MF (*top row*) and cat-MF (*bottom row*).The highest concentration was placed in the tubes on the *left hand side*. TR: 3500 ms, TE: 10 ms. **c** Representative inverse-recovery spin echo image of the same MF and cat MF concentration range. TR: 6000 ms, TE: 9.6 ms

values for MF cores of similar size containing iron oxide only [6]. This may be due to the addition of 1% cobalt to MF in this work and a resulting increase in magnetic susceptibility. Interestingly, r2 of cat-MF was found to be 18% higher compared to MF. Surface functionalisation can affect r2 by influencing the mobility of surrounding water molecules [20]. However, amine-functionalised SPIONs have been shown to actually slightly reduce r2 compared to carboxyl-functionalised SPIONs, or SPIONs displaying a mixture of amine and carboxyl groups on their surface [14]. Therefore, the reason for the apparent increase in r2 of cat-MF remains unclear.

3.4 Conclusions and Future Directions

It has been shown that MF could be successfully synthesised from ApoF and cationised using established protocols. Cationisation efficiency was comparable to previously reported efficiencies using ferritin, indicating that the protein shell was not significantly altered by MF synthesis. MALDI-TOF measurements showed that not all acidic residues present in MF were cationised. Future work using multiple additions of DMPA and EDC would elucidate whether higher cationisation efficiencies could be achieved. If multiple additions do not result in higher cationisation efficiencies, it could be concluded that all solvent accessible acidic residues have been cationised and that the remaining aspartic and glutamic acid residues are "buried" within the protein subunits and not accessible.

The magnetic properties of cat-MF were found to be similar to those of MF, showing that cationisation of the protein shell did not affect the magnetic properties of the core. However, r2 was higher for cat-MF compared to MF. Further measurements on a variety of cat-MF batches would be needed to establish whether this is a reproducible result or within the measurement error.

References

1. Danon, D., Skutelsk. E., Marikovs, Y., Goldstei, L.: Use of cationized ferritin as a label of negative charges on cell surfaces. J. Ultrastruct. Res. **38** (5–6), 500–510 (1972)
2. Perriman, A.W., Cölfen, H., Hughes, R.W., Barrie, C.L., Mann, S.: Solvent-Free protein liquids and liquid crystals. Angew. Chem. Int. Ed. **48**(34), 6242–6246 (2009)
3. Klem, M.T., Resnick, D.A., Gilmore, K., Young, M., Idzerda, Y.U., Douglas, T.: Synthetic control over magnetic moment and exchange bias in all-oxide materials encapsulated within a spherical protein cage. J. Am. Chem. Soc. **129**(1), 197–201 (2007)
4. Meldrum, F.C., Heywood, B.R., Mann, S.: Magnetoferritin: in vitro synthesis of a novel magnetic protein. Science **257**(5069), 522–523 (1992)
5. Okuda, M., Eloi, J.C., Sarua, A., Jones, S.E.W., Schwarzacher, W.: Energy barrier distribution for dispersed mixed oxide magnetic nanoparticles. J. Appl. Phys. **111**, 7(2012)
6. Uchida, M., et al.: A human ferritin iron oxide nano-composite magnetic resonance contrast agent. Magn. Reson. Med. **60**(5), 1073–1081 (2008)
7. Arosio, P., Adelman, T.G., Drysdale, J.W.: On ferritin heterogeneity. further evidence for heteropolymers. J. Biol. Chem. **253**(12), 4451–4458 (1978)
8. Bean, C., Livingston, J.: Superparamagnetism. J. Appl. Phys. **30**(4), S120–S129 (1959)
9. Lu, A.H., Salabas, E.L., Schüth, F.: Magnetic nanoparticles: synthesis, protection, functionalization, and application. Angew. Chem. Int. Ed. **46**(8), 1222–1244 (2007)
10. Kim, D., Zhang, Y., Voit, W., Rao, K., Muhammed, M.: Synthesis and characterization of surfactant-coated superparamagnetic monodispersed iron oxide nanoparticles. J. Magn. Magn. Mater. **225**(1), 30–36 (2001)
11. Woo, K., et al.: Easy synthesis and magnetic properties of iron oxide nanoparticles. Chem. Mater. **16**(14), 2814–2818 (2004)
12. Slonczewski, J.: Origin of magnetic anisotropy in cobalt-substituted magnetite. Phys. Rev. **110**(6), 1341 (1958)
13. Stanisz, G.J., et al.: T1, T2 relaxation and magnetization transfer in tissue at 3t. Magn. Reson. Med. **54**(3), 507–512 (2005)
14. Tong, S., Hou, S., Zheng, Z., Zhou, J., Bao, G.: Coating optimization of superparamagnetic iron oxide nanoparticles for high T2 relaxivity. Nano Lett. **10**(11), 4607–4613 (2010)
15. Tromsdorf, U.I., et al.: Size and surface effects on the MRI relaxivity of manganese ferrite nanoparticle contrast agents. Nano Lett. **7**(8), 2422–2427 (2007)
16. Macher, T., et al.: Ultrathin iron oxide nanowhiskers as positive contrast agents for magnetic resonance imaging. Adv. Func. Mater. **25**(3), 490–494 (2015)
17. Na, H.B., Song, I.C., Hyeon, T.: Inorganic nanoparticles for MRI contrast agents. Adv. Mater. **21**(21), 2133–2148 (2009)

18. Wang, Y.X., Hussain, S.M., Krestin, G.P.: Superparamagnetic iron oxide contrast agents: physicochemical characteristics and applications in MR imaging. Eur. Radiol. **11**(11), 2319–2331 (2001)
19. Simon, G.H., et al.: T1 and T2 relaxivity of intracellular and extracellular uspio at 1.5 t and 3t clinical mr scanning. Eur. Radiol. **16**(3), 738–745 (2006)
20. Duan, H., Kuang, M., Wang, X., Wang, Y.A., Mao, H., Nie, S.: Reexamining the effects of particle size and surface chemistry on the magnetic properties of iron oxide nanocrystals: new insights into spin disorder and proton relaxivity. J. Phys. Chem. C **112**(22), 8127–8131 (2008)

Chapter 4
Stem Cell Labelling with Cationised Magnetoferritin

Magnetic labelling of stem cells with SPIONs has many important applications, such as remote manipulation [1] or imaging [2]. To achieve adequate magnetisation for these applications, sufficient cellular uptake of SPIONs is necessary. Therefore, SPIONs have been functionalised with a variety of molecular species with the aim to improve cell internalisation [3–5] Ideally, short incubation times and low extracellular iron dose should be combined to avoid potential adverse effects [6, 7], but at the same time sufficient intracellular iron levels must be achieved under these conditions. In this work, cationisation of MF was explored as a route to create a novel SPION that can rapidly magnetise cells using low exposure concentrations and incubation times as short as 1 min. In this chapter, the magnetisation efficiency of cat-MF is explored in human mesenchymal stem cells (hMSCs) using MACS and MRI. First, the time- and concentration-dependence of the magnetic labelling process is investigated, followed by a study into long-term retention of magnetisation up to five weeks after labelling. Finally, the results of a mechanistic study investigating whether cell magnetisation is mediated by electrostatic adsorption of cat-MF to anionic domains on the cell membrane are presented.

4.1 Magnetisation Efficiency: Time and Concentration Dependence

Magnetisation efficiency was determined as a function of incubation time and extracellular cat-MF concentration. First, cells were incubated with 0.5 μM of cat-MF for 1 min to 6 h at 37 °C. The percentage of magnetised cells and the cellular iron content measured in the magnetised cell fraction for the first 60 min of labelling are shown in Fig. 4.1a. It was found that cat-MF was remarkably effective at magnetically labelling hMSCs. Notably, a 1 min exposure resulted in the magnetisation of 92% of the cell population and the delivery of approximately 3.6 pg of

S. Correia Carreira, *Rapid Cell Magnetisation Using Cationised Magnetoferritin*,
Springer Theses, DOI 10.1007/978-3-319-60333-9_4

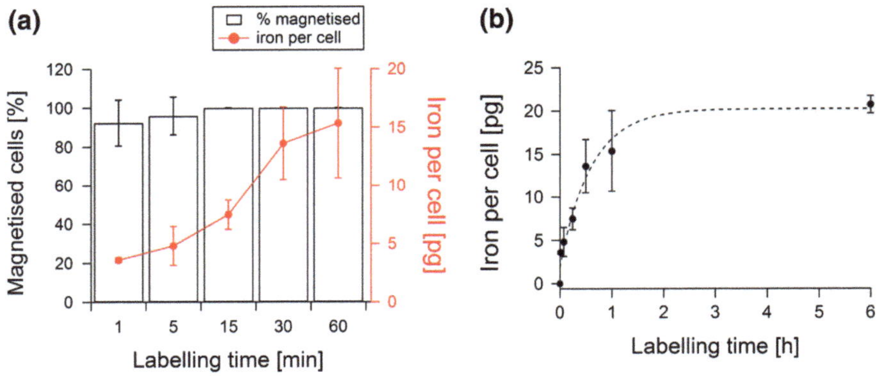

Fig. 4.1 MACS and ICP-OES analysis of hMSC as a function of labelling time. Labelling concentration was 0.5 μM cat-MF. Average and standard deviation from three biological replicates (i.e. hMSCs derived from three patients) are shown. **a** Percentage of magnetised hMSCs and iron content per cell measured in the magnetised cell fraction up to a labelling time of 1 h. 92% of the cell population were magnetised after a 1 min exposure, and the entire cell population was magnetised after 15 min. **b** Iron content per cell over 6 h incubation. Saturation of iron uptake was observed after 6 h incubation with a maximum cellular iron content of 20.2 pg/cell. *Dashed line* exponential fit to the experimental data

iron per cell, which corresponds to roughly 1×10^8 nanoparticles per cell. Remarkably, given the brevity of exposure, this cellular iron content is within the reported range required to influence T2 contrast for MRI [8, 9]. Increasing the incubation time to 30 min resulted in the magnetisation of the entire cell population and a cellular iron content of approximately 13 pg $cell^{-1}$. Figure 4.1b shows that iron uptake into cells was saturated after a 6 h incubation. The exponential fit to the data revealed a maximum cellular iron content of 20.2 pg $cell^{-1}$ and a time constant of 0.6 h. Thus, it took just over 30 min for the cells to reach 63% of the maximum iron loading. The uptake rate between one and 30 min was 20.5 pg $cell^{-1}$ h^{-1}, however, between 30 min and 6 h this rate slowed to 1.2 pg $cell^{-1}$ h^{-1}.

Next, cells were incubated with cat-MF concentrations ranging from 0.01 to 2 μM for 30 min at 37°. It was found that the magnetisation efficiency depended on incubation concentration, with more concentrated solutions of cat-MF magnetising a greater proportion of hMSCs and delivering a higher cellular iron content (Fig. 4.2a).

For the interpretation of these results, it is useful to calculate the minimum number of cat-MF particles needed to magnetise hMSCs sufficiently for retention in the MACS column. The magnetic force acting on the labelled cell needs to compensate the viscous drag force in order to immobilise the cells in the column. Given that the cells are loaded onto the column in a volume of 0.5 mL and that this volume moves through the magnetised part of the column (20 mm) within 1 min, the velocity of the cells is 3.3×10^{-4} ms^{-1}. Assuming that the cell has a radius of 5 μm, the resulting drag force is 3×10^{-11} N (see Eq. 2.15). Substituting this value into Eq. 18, it follows that the minimum number of cat-MF particles needed to

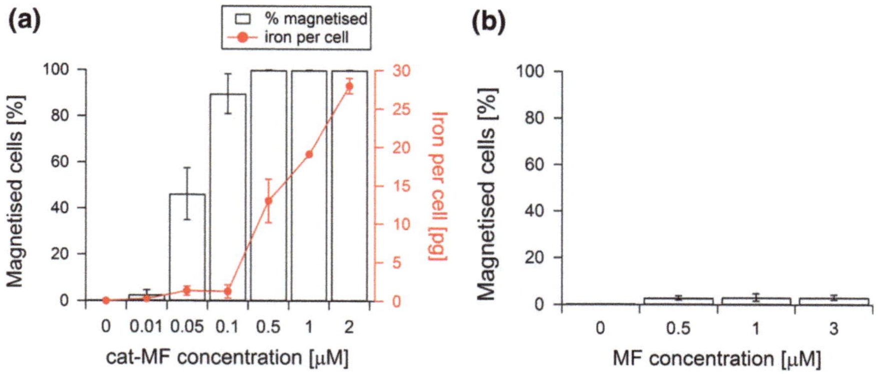

Fig. 4.2 MACS and ICP-OES analysis of hMSC as a function of extracellular nanoparticle concentration. Labelling time was 30 min. Average and standard deviation from three biological replicates are shown. **a** Magnetisation efficiency and cellular iron content after labelling with cat-MF. Exposure to 0.5 μM cat-MF results in the magnetisation of the entire cell population. **b** Magnetisation efficiency of hMSC exposed to un-functionalised MF. Iron content was below the limit of ICP-OES

magnetise the cells sufficiently is 1.5×10^7. Results in Fig. 4.2a show that an exposure concentration of at least 0.05 μM cat-MF was required to magnetise 50% of hMSCs. Exposure to 0.05 μM cat-MF resulted in a cellular iron content of 1.3 pg, or 3.8×10^7 cat-MF particles per cell. This value agrees well with the theoretical prediction that a minimum of 1.5×10^7 cat-MF particles would be required to overcome the viscous drag force that the cells experience when they pass through the column. The fact that only 50% of the hMSCs were magnetised using this cat-MF concentration, indicates that the other half of the cell population did not incorporate sufficient magnetic nanoparticles from the available cat-MF solution. This could be due to a combination of the low number of nanoparticles available in the diluted cat-MF solution and variations in the core sizes (and thus iron loading) inside cat-MF.

The discrepancy between cellular iron content and percentage magnetisation observed in Fig. 4.2a is an interesting phenomenon. It is shown that a 30 min incubation with both 0.05 and 0.1 μM cat-MF yielded an average cellular iron content of 1.3 pg, but the lower cat-MF dose only magnetised 45% of the cell population, while the higher dose magnetised 90%. A possible explanation for this discrepancy in percentage magnetisation observed in populations that contain similar amounts of cellular iron could be the number and therefore availability of cat-MF particles in the two solutions. It could be that only a small number of hMSCs came into contact with a sufficient number of cat-MF particles when they were exposed to the dilute 0.05 μM cat-MF solution, resulting in the low magnetisation efficiency observed. However, when exposed to 0.1 μM cat-MF, it became more likely that the sufficient number of particles reached almost all hMSCs in the cell layer, because of the increased particle concentration. As a result, many more cells contained the required cellular iron content and were thus magnetised.

The minimal cat-MF concentration needed to magnetise the entire cell population was 0.5 μM. In contrast, incubation of hMSC with 3 μM un-functionalised MF magnetised less than 3% of the cell population and cellular iron content was below the ICP-OES detection limit (Fig. 4.2b). This result demonstrates the direct impact of cationisation on the efficiency of the magnetic labelling process, and is in agreement with previous studies showing that native ferritin is taken up in very small amounts, whilst cationised ferritin binds rapidly to the cell membrane and is found within vesicles inside the cell within 15 to 60 min of exposure [10].

Other charge functionalisations are likely to achieve similarly rapid magnetisation as described here, however, so far no study has demonstrated that effective magnetic labelling can in fact be achieved using incubation times well below 30 min. For example, it has been shown that anionic nanoparticles were internalised very efficiently and iron levels of 10 pg per cell could be achieved within a 30 min incubation period with 5 mM iron [11]. In comparison, incubation with a cat-MF solution containing 0.5 μM protein corresponds to incubation with approximately 0.2 mM iron and also yields approximately 10 pg of iron per cell after 30 min. Using an extracellular iron concentration similar to the one presented here, a 1 h incubation period with TAT-peptide functionalised SPIONs achieved cellular iron loading of approximately 18 pg per cell in neural progenitor cells [4], which is similar to the 15 pg achieved here after 1 h incubation with cat-MF. Taken together, the results presented in this work show for the first time that efficient magnetic labelling can be achieved within a 1 min incubation period, and that this efficiency can also be achieved using relatively low extracellular iron concentrations.

4.2 MRI of hMSCs Labelled with MF and cat-MF

The suitability of cat-MF as a cellular MRI contrast agent was investigated by imaging unlabelled hMSCs, as well as hMSCs labelled with MF and cat-MF using a 3 T MRI scanner. Labelling hMSCs with 0.5 or 1 μM un-functionalised MF for 30 min did not affect $T1^{-1}$ or $T2^{-1}$ compared to unlabelled cells. However, $T2^{-1}$ was increased by up to 2.5 times after incubation with equivalent amounts of cat-MF (Fig. 4.3a). A gradient echo sequence was chosen for imaging, because it yields images whose contrast is particularly sensitive to field inhomogeneities, and thus the presence of SPIONs [12]. Little contrast enhancement was observed in cells exposed to MF, however, significant signal loss was detected in the area containing hMSCs labelled with cat-MF (Fig. 4.3b).

T1 and T2 shortening was much reduced once cat-MF was immobilised inside hMSCs compared to cat-MF in solution. Taking into account the amount of iron measured in the cat-MF labelled hMSCs, cellular r1 and r2 in cat-MF labelled cells was 0.2 and 7.7 mM^{-1} s^{-1}, respectively. This is a significant reduction compared to the r1 and r2 values of 2.3 and 52.8 mM^{-1} s^{-1}, respectively, measured for cat-MF

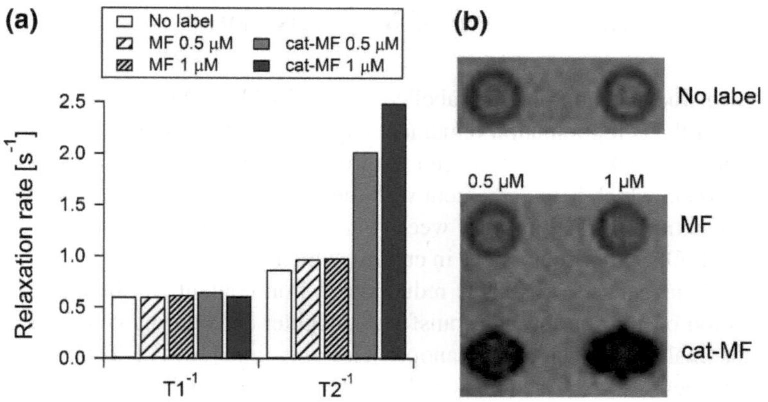

Fig. 4.3 MRI on 750,000 hMSCs exposed to two concentrations of MF and cat-MF for 30 min. **a** Little effect on $T1^{-1}$ was observed upon cellular internalisation of MF and cat-MF. Labelling with cat-MF led to a significant increase of $T2^{-1}$, but MF uptake was not sufficient to affect $T2^{-1}$ compared to unlabelled cells. **b** Gradient echo image (TR = 100 ms, TE = 10 ms, flip angle 30°) of unlabelled hMCs, and hMSC exposed to two concentrations of MF and cat-MF

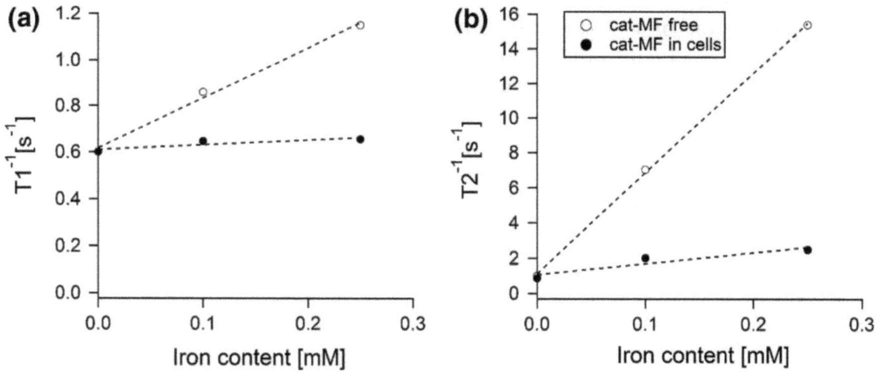

Fig. 4.4 Effect of cellular internalisation of cat-MF on relaxation rates. **a** The longitudinal and **b** the transverse relaxation rate as a function of iron content in the sample. The *dashed line* is the linear fit to the experimental data and represents the molar relaxivity: $r1_{free}$: 2.3 $mM^{-1}s^{-1}$, $r1_{cell}$: 0.2 $mM^{-1}s^{-1}$, $r2_{free}$: 52.8 $mM^{-1}s^{-1}$, $r2_{cell}$: 7.7 $mM^{-1}s^{-1}$

in solution (Fig. 4.4; compare also Table 3.2). A marked decrease in relaxivites has previously been observed in SPIONs internalised by cells, which is due to limited water and/or SPION diffusion inside endocytotic vesicles [3, 13]. Here, r1 is reduced ten-fold and r2 approximately sevenfold after cell uptake. Similarly, it has been shown that r1 is affected to a greater extent than r2, presumably because spin-spin interactions that affect T2 can still take place whilst spin-lattice interactions of SPIONs are significantly inhibited due to the particles' confinement [14].

4.3 Long-Term Retention of Magnetisation

Long-term studies using hMSCs labelled with 0.5 μM cat-MF for 30 min revealed that 60% of the cell population remained magnetised after one week in culture, and 10% after five weeks (Fig. 4.5a). Cellular iron content decreased by approximately 60% per week, which is in agreement with the observation of a population doubling in the magnetic cell fraction every week (Fig. 4.5b). It dropped below the detection limit of ICP-OES after four weeks in culture, which is accompanied by a drop in the number of magnetised cells. This reduction in iron content can be attributed to a combination of dilution through transfer to daughter cells during cell division [15], and lysosomal breakdown of the nanoparticles over time [16, 17]. Total iron content in the magnetised cell fraction is reduced by approximately 9% during the first week, and drops by 39% between the first and second week (Fig. 4.5b). Five weeks after labelling the iron content per cell is below the limit of detection for ICP-OES, however, 7% of the total cell population are still magnetised sufficiently to be retained in the MACS column. It has previously been shown that SPION-labelled MSC retained magnetisation up to 12 weeks in vivo [18], and that some SPIONs sequestered by macrophages remained intact while coexisting with degraded SPIONs inside lysosomes up to two months in vivo [19]. Therefore, it is possible that a proportion of cat-MF still remained intact inside hMSCs after 5 weeks, retaining their magnetic order. The proposed mechanism for SPION biotransformation suggests that internalised SPIONs are eventually located in lysosomes, in

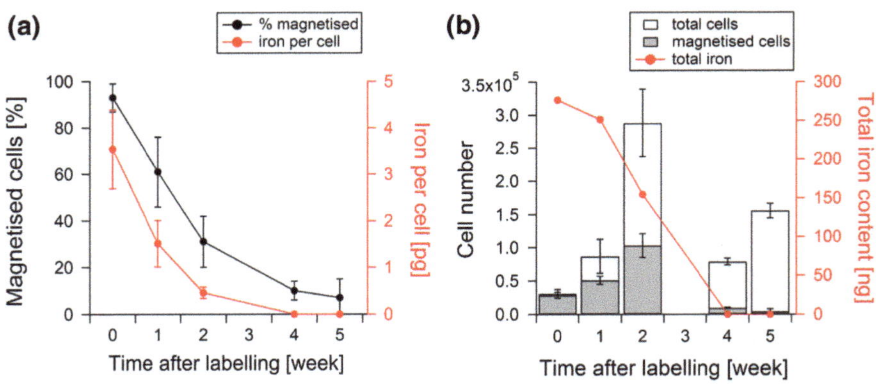

Fig. 4.5 Retention of cellular magnetisation over five weeks after labelling with 0.5 μM cat-MF for 30 min. **a** MACS and ICP-OES analysis of hMSCs over five weeks. **b** Total cell number, cell number in the magnetised fraction and total iron content in the magnetised fraction over five weeks. Note that the overall cell number appears decreased after three weeks, because cells had reached confluency and were sub-cultured after three weeks. Average and standard deviation of three biological replicates are shown

which they are degraded. The released iron is then stored in intracellular ferritin [19]. Whether this mechanism is the same for a SPION already encapsulated in ferritin, as is the case here, is unknown. Equally, it would be interesting to study whether the introduced ferritin cages would be "recycled" as iron-storage vessels by hMSCs, seeing that the ferritin used here stems from a different species (horse).

Another interesting point concerns the loss of magnetisation due to dilution of cat-MF through transfer to daughter cells during cell division, as was apparent in the increase in the number of cells in the magnetised fraction over the first two weeks (Fig. 4.5b). It is possible that magnetisation could have been retained for longer if hMSCs had been kept from dividing, e.g. by exposing them to cat-MF when they had already grown to confluency. In fact, it has been shown that confluent hMSC labelled with SPIONs retained approximately 80% of the cellular iron content and the ability to significantly shorten T2 up to six weeks after labelling [20]. This would suggest that the primary cause of magnetisation loss observed here may have been due to dilution of cat-MF as a result of cell proliferation.

MRI on hMSCs labelled with cat-MF showed that contrast enhancement persisted for at least one week (Fig. 4.6). Although $T2^{-1}$ was decreased by 30% compared to the first day of labelling, it was still almost twice as high compared to regions containing unlabelled hMSCs. This indicates that cat-MF could be used to track stem cells over time using MRI. As shown above, cellular iron content likely decreased over time because of cat-MF dilution as a result of cell division. It is difficult to predict how long cat-MF labelled hMSCs would stay magnetised after in vivo injection, because the cell proliferation rate may be different compared to in vitro culture.

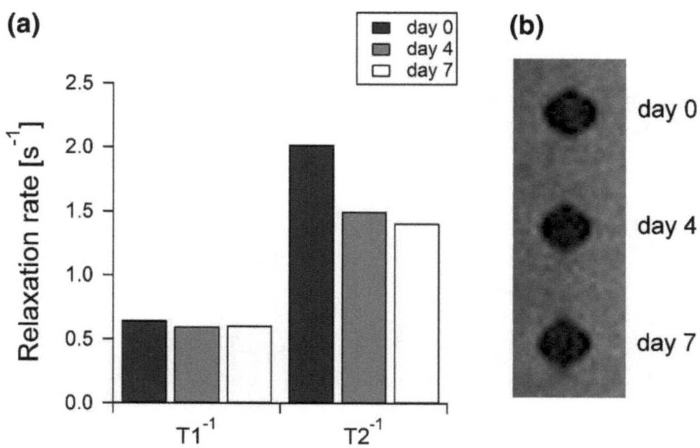

Fig. 4.6 MRI of 750,000 hMSCs up to one week after labelling with cat-MF. **a** Little effect on $T1^{-1}$ is observed over time, whilst $T2^{-1}$ is decreased by 30% after one week. **b** Gradient echo image (TR = 100 ms, TE = 10 ms, flip angle 30°) of hMSC labelled with cat-MF up to 7 days after labelling

4.4 TEM and Light Microscopy of hMSCs Labelled with cat-MF

TEM of magnetised hMSCs revealed widespread coverage of the cell surface with cat-MF immediately after 30 min of labelling (Fig. 4.7a), but not after the magnetised hMSCs had been cultured for one week (Fig. 4.7b).

Given that the majority of cells was still magnetised at this time point, internalisation of cat-MF must have occurred. However, intracellular cat-MF could not be confirmed with confidence using TEM imaging, and no endocytotic vesicles, such as clathrin-coated or uncoated vesicles could be identified. The difficulty in observing such cellular ultrastructure in Fig. 4.7 may have been a result of the excessively heavy stain, which limited the conclusions that could be drawn from the TEM images. Therefore, the uptake mechanism is unknown. However, Prussian Blue staining of sectioned hMSC 48 h after exposure to 0.5 μM cat-MF showed that the stain was present inside the cells, providing evidence that cat-MF had indeed been internalised (Fig. 4.8a). The blue stain was not visible in sections of untreated hMSCs (Fig. 4.8b).

Previous TEM-based studies investigating the internalisation of cationised ferritin have indicated that internalisation occurs within the first 10 min of exposure, with cationised ferritin being localised in coated vesicles [21, 22]. This indicates either clathrin- or caveolin-dependent endocytosis. These studies also reported that after 30 min incubation, cationised ferritin was still present on the cell surface and could also be localised in multivesicular bodies, some of them resembling lysosomes.

4.5 Elucidating the Mechanism Underlying Rapid Labelling

The incubation conditions were systematically varied to understand the different contributions of electrostatic interaction and internalisation in cat-MF labelling (Fig. 4.9). Labelling at 37 and 4 °C for 5 min resulted in similar cellular iron

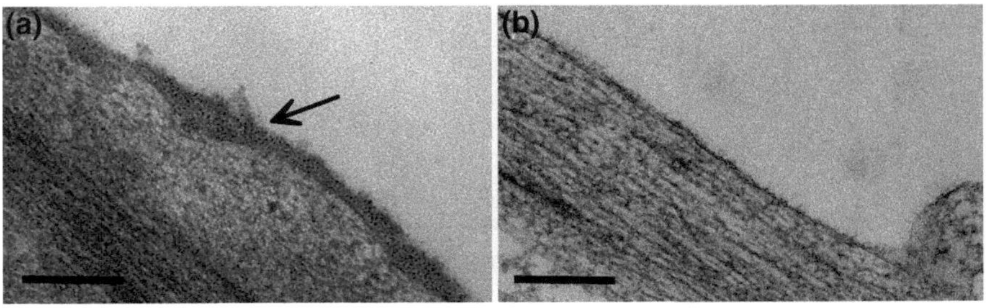

Fig. 4.7 TEM of hMSCs exposed to cat-MF. **a** Immediately after exposure to 0.5 μM cat-MF for 30 min. The *arrow* indicates cat-MF particles densely packed on the cell surface. **b** One week after exposure. No cat-MF is present on the cell surface any more. *Scale bar* 200 nm

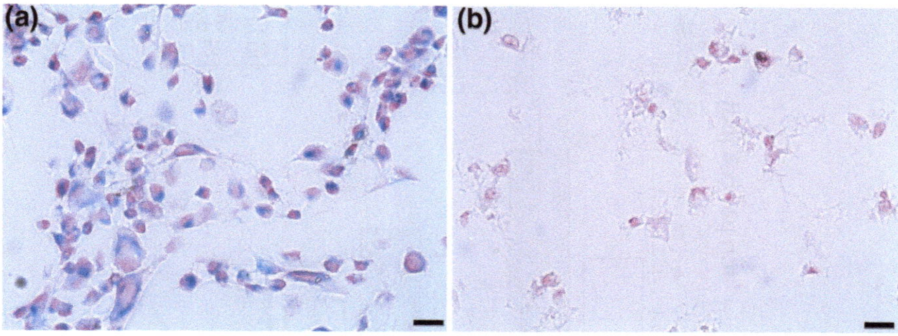

Fig. 4.8 Prussian Blue staining of sectioned hMSCs. **a** 48 h after exposure to cat-MF. The *blue stain* is localised inside the cells and indicates that cat-MF has been internalised. Cell nuclei were counter-stained with Nuclear Fast Red and appear *pink*. **b** Untreated hMSCs. No blue stain is visible. *Scale bar* 20 μm

content, however, after 30 min the iron content was significantly higher in cells labelled at 37 °C compared to cells labelled at 4 °C. Low temperatures are an efficient method of non-invasively inhibiting many active internalisation pathways [23]. Thus, the iron content measured at 4 °C is likely to represent cat-MF bound to the cell surface rather than internalised nanoparticles, and iron content measured at 37° represents the adsorbed as well as the internalised fraction of cat-MF. These results demonstrate that active internalisation pathways contribute to increased cellular iron content as labelling time progresses.

Culturing hMSCs in medium supplemented with 80 mM sodium chlorate ($NaClO_3$) significantly reduced the cellular iron content after a 5 min incubation with cat-MF (Fig. 4.9). $NaClO_3$ inhibits the synthesis of anionic functional groups on proteoglycans [25], which suggests that these glycocalyx species act as binding sites for cat-MF during the initial labelling period, as has been demonstrated for other cationic ligands [26]. The observation that $NaClO_3$-treated cells still contained a relatively large amount of iron suggests that other anionic moieties, such as phospholipid head groups, may also contribute to electrostatic binding. After 30 min, the cellular iron content in cells cultured in $NaClO_3$ was still much reduced compared to cells cultured in untreated medium. This indicates that initial electrostatic adsorption of cat-MF to the cell surface is important for high internalisation rates, which is in agreement with previously calculated models that predict a higher rate of endocytosis with increased ligand-receptor interactions [27]. These results are also consistent with models showing that non-specific interactions are as important as specific interactions during endocytosis, and favour nanoparticle uptake [28].

Additionally, previous studies have shown that anionised and cationised ferritin could be used as a TEM stain for charged functional groups on the cell surface [29]. It was shown that cationised ferritin attached uniformly to the plasma membrane of cells, demonstrating that the cell surface possesses a similarly uniform distribution

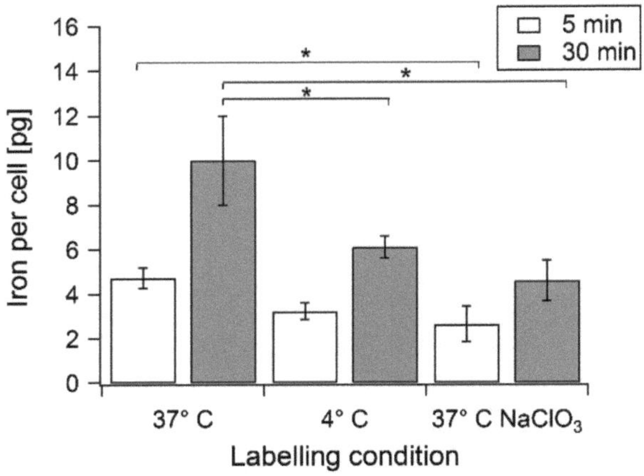

Fig. 4.9 Mechanistic study of cat-MF uptake. ICP-OES was used to analyse the cellular iron content of hMSCs incubated with 0.5 μM cat-MF for either 5 or 30 min at 37, 4, or 37 °C after $NaClO_3$ treatment. Average and standard deviation of three biological replicates are shown. For each incubation period, cellular iron content values were compared using the Kruskal-Wallis test and *post hoc* paired comparison, with significant differences indicated with an asterisk ($p < 0.05$). Image re-printed with permission from Correia Carreira et al. [24]

of anionic domains to which the cationised protein can bind [29, 30]. Native or anionic ferritin on the other hand attached much more scarcely and in patches to the cell membrane, and was also internalised less efficiently [10, 30].

4.6 Conclusions and Future Directions

Efficient magnetisation of hMSCs was achieved with cat-MF using incubation times as short as 1 min with an extracellular iron concentration of only 10 μg mL^{-1} (0.2 mM). This result represents the first report of effective cell labelling using incubation times well below 30 min in conjunction with low dose magnetic nanoparticle exposure. Un-functionalised MF did not achieve efficient magnetisation within a 30 min incubation period even at elevated extracellular concentrations, demonstrating the significance of cationisation for rapid and efficient magnetic labelling. Furthermore, hMSCs labelled with cat-MF were retained in MACS columns, and induced significant signal intensity loss in gradient echo MR images for at least one week after labelling. Finally, it was shown that rapid magnetic labelling was mediated by adsorption of cat-MF to anionic functional groups in the glycocalyx.

The exact uptake mechanism was not elucidated in this work, but two facts could be established: cat-MF uptake was saturable (Fig. 4.1b) and cat-MF binds to surface proteoglycans during the first few minutes of labelling (Fig. 4.9). The observation that cat-MF uptake was saturable indicates that the uptake mechanism is

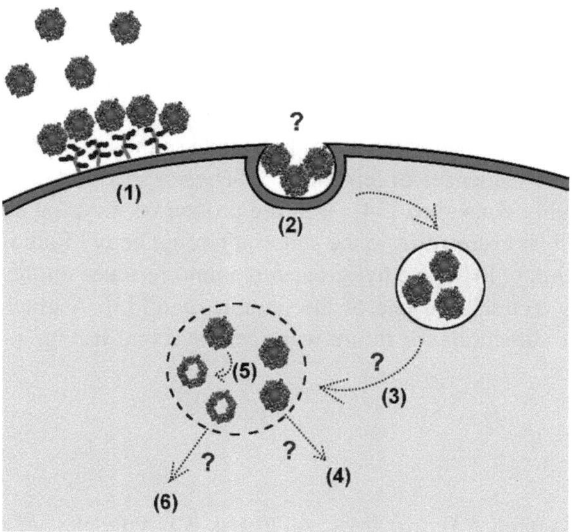

Fig. 4.10 Conclusions and directions for future work. Magnetic labelling with cat-MF has shown to be very rapid and mediated by electrostatic adsorption of the cationised nanoparticle to anionic domains on the cell surface, such as proteoglycans (*1*). The exact internalisation mechanism is unknown (*2*), as well as the fate of cat-MF in the cell. If cat-MF reaches the lysosomal compartment (*3*), several scenarios are possible. The nanoparticles could escape the lysosome due to the proton-sponge effect (*4*), or the iron core could be degraded by lysosomal hydrolases (*5*). Finally, the fate of the protein cage is unknown, and it would be interesting to investigate whether it is recycled for iron storage (*6*)

dependent on ligands or auxiliary proteins that enable the internalisation of cat-MF. Payne et al. [26]. have shown that cationic polypeptides were internalised after binding to cell surface proteoglycans via a clathrin- and caveolin-independent, but flotillin- and dynamin-dependent internalisation pathway. The authors showed that proteoglycan-bound ligands associated with flotillin-1-positive vesicles and were trafficked to late endosomes. Early TEM-based studies showed that cationised ferritin was internalised via coated pits [21, 22, 30], and given the results from Payne et al. those pits could have been coated with flotillin proteins. Detailed studies, in which the cellular iron content is measured after administration of endocytotic blockers could help elucidate the uptake mechanism. Furthermore, the fate of cat-MF, once internalised in hMSCs, could be explored by fluorescently labelling cat-MF and incubation of exposed cells with fluorescent dyes that stain the acidic compartment of lysososmes. Thus, co-localisation of both fluorescent signals would indicate if and when cat-MF is present in lysosomes. There has been some evidence that cationic nanoparticles can escape lysosomes due to the so-called proton sponge effect [31]. Unsaturated amine groups on the cationic particle surface can sequester the protons that are supplied through the lysosomal proton pump. As a consequence, chloride ions are retained in the lysosome and lead to an influx of water, which ultimately causes swelling and rupture of the vesicle. It is unclear if cat-MF could have the same effect, seeing that the zeta potential (and therefore the cationic charge density) is relatively low compared to, for example, polycationic polymers that are used to induce the proton-sponge effect to aid the escape of DNA from the lysosome in gene transfection experiments [32, 33].

Finally, it would be interesting to study the fate of the introduced ferritin protein cage. Amino acid sequences of ferritin vary between species but the structure and function are highly conserved [34]. Will the protein be recycled and used for iron storage or will it be excreted from the cell as a foreign body? Radio-labelling of the ferritin by attaching [14] C-methyl groups to amino residues on the protein surface could be a way to track the fate of the protein cage [35]. A graphic summary of conclusions and directions for future work can be found in Fig. 4.10.

References

1. Dobson, J., Cartmell, S.H., Keramane, A., El Haj, A.J.: Principles and design of a novel magnetic force mechanical conditioning bioreactor for tissue engineering, stem cell conditioning, and dynamic in vitro screening. IEEE Trans. Nanobiosci. **5**(3), 173–177 (2006)
2. Mahmoudi, M., et al.: Magnetic resonance imaging tracking of stem cells in vivo using iron oxide nanoparticles as a tool for the advancement of clinical regenerative medicine. Chem. Rev. **111**(2), 253–280 (2010)
3. Bulte, J.W.M., et al.: Magnetodendrimers allow endosomal magnetic labeling and in vivo tracking of stem cells. Nat. Biotechnol. **19**(12), 1141–1147 (2001)
4. Lewin, M., et al.: Tat peptide-derivatized magnetic nanoparticles allow in vivo tracking and recovery of progenitor cells. Nat. Biotechnol. **18**(4), 410–414 (2000)
5. Wilhelm, C., Billotey, C., Roger, J., Pons, J., Bacri, J.-C., Gazeau, F.: Intracellular uptake of anionic superparamagnetic nanoparticles as a function of their surface coating. Biomaterials **24**(6), 1001–1011 (2003)
6. Daldrup-Link, H.E., et al.: Targeting of hematopoietic progenitor cells with mr contrast agents. Radiology **228**, 760–767 (2003)
7. Singh, N., Jenkins, G.J., Asadi, R., Doak, S.H.: Potential toxicity of superparamagnetic iron oxide nanoparticles (Spion). Nano Rev. **1,** (2010)
8. Smirnov, P., et al.: In vivo cellular imaging of lymphocyte trafficking by MRI: a tumor model approach to cell-based anticancer therapy. Magn. Reson. Med. **56**(3), 498–508 (2006)
9. Billotey, C., et al.: T-Cell homing to the pancreas in autoimmune mouse models of diabetes: in vivo MR Imaging. Radiology. **236**(2), 579–587 (2005)
10. Farquhar, M.G.: Recovery of surface membrane in anterior pituitary cells. variations in traffic detected with anionic and cationic ferritin. J. Cell Biol. **77**(3), R35–R42 (1978)
11. Wilhelm, C., et al.: Magnetic control of vascular network formation with magnetically labeled endothelial progenitor cells. Biomaterials **28**(26), 3797–3806 (2007)
12. Stark, D.D., et al.: Superparamagnetic iron oxide: clinical application as a contrast agent for MR imaging of the liver. Radiology **168**(2), 297–301 (1988)
13. Simon, G.H., et al.: T1 and T2 relaxivity of intracellular and extracellular Uspio at 1.5 T and 3t clinical Mr scanning. Eur. Radiol. **16**(3), 738–745 (2006)
14. Billotey, C., Wilhelm, C., Devaud, M., Bacri, J.C., Bittoun, J., Gazeau, F.: Cell internalization of anionic maghemite nanoparticles: quantitative effect on magnetic resonance imaging. Magn. Reson. Med. **49**(4), 646–654 (2003)
15. Wilhelm, C., Gazeau, F.: Universal cell labelling with anionic magnetic nanoparticles. Biomaterials **29**(22), 3161–3174 (2008)

16. Arbab, A.S., Wilson, L.B., Ashari, P., Jordan, E.K., Lewis, B.K., Frank, J.A.: A model of lysosomal metabolism of dextran coated superparamagnetic iron oxide (Spio) nanoparticles: implications for cellular Magnetic Resonance Imaging. NMR Biomed. **18**(6), 383–389 (2005)
17. Soenen, S.J., Himmelreich, U., Nuytten, N., Pisanic, T.R., Ferrari, A., De Cuyper, M.: Intracellular nanoparticle coating stability determines nanoparticle diagnostics efficacy and cell functionality. Small **6**(19), 2136–2145 (2010)
18. Jing, X.H., et al.: In vivo MR imaging tracking of magnetic iron oxide nanoparticle labeled, engineered, autologous bone marrow mesenchymal stem cells following intra-articular injection. Joint Bone Spine. **75**(4), 432–438 (2008)
19. Levy, M., et al.: Long term in vivo biotransformation of iron oxide nanoparticles. Biomaterials **32**(16), 3988–3999 (2011)
20. Arbab, A.S., et al.: Characterization of biophysical and metabolic properties of cells labeled with superparamagnetic iron oxide nanoparticles and transfection agent for cellular MR imaging. Radiology **229**(3), 838–846 (2003)
21. MacLean, I., Sanders, E.: Cationized ferritin and phosvitin uptake by coated vesicles of the early chick embryo. Anat. Embryol. **166**(3), 385–397 (1983)
22. Van Deurs, B., Nilausen, K., Faergeman, O., Meinertz, H.: Coated pits and pinocytosis of cationized ferritin in human skin fibroblasts. Eur. J. Cell Biol. **27**(2), 270–278 (1982)
23. Steinman, R.M., Mellman, I.S., Muller, W.A., Cohn, Z.A.: Endocytosis and the recycling of plasma-membrane. J. Cell Biol. **96**(1), 1–27 (1983)
24. Carreira, S.C., Armstrong, J., Seddon, A., Perriman, A., Hartley-Davies, R., Schwarzacher, W.: Ultra-fast stem cell labelling using cationised magnetoferritin. Nanoscale **8**(14), 7474–7483 (2016)
25. Baeuerle, P.A., Huttner, W.B.: Chlorate–a potent inhibitor of protein sulfation in intact cells. Biochem. Biophys. Res. Commun. **141**(2), 870–877 (1986)
26. Payne, C.K., Jones, S.A., Chen, C., Zhuang, X.: Internalization and trafficking of cell surface proteoglycans and proteoglycan-binding ligands. Traffic **8**(4), 389–401 (2007)
27. Vacha, R., Martinez-Veracoechea, F.J., Frenkel, D.: Receptor-mediated endocytosis of nanoparticles of various shapes. Nano Lett. **11**(12), 5391–5395 (2011)
28. Decuzzi, P., Ferrari, M.: The role of specific and non-specific interactions in receptor-mediated endocytosis of nanoparticles. Biomaterials **28**(18), 2915–2922 (2007)
29. Burry, R.W., Wood, J.G.: Contributions of lipids and proteins to the surface charge of membranes. An electron microscopy study with cationized and anionized ferritin. J. Cell Biol. **82**(3), 726–741 (1979)
30. Mutsaers, S.E., Papadimitriou, J.M.: Surface charge of macrophages and their interaction with charged particles. J. Leukoc. Biol. **44**(1), 17–26 (1988)
31. Nel, A.E., et al.: Understanding Biophysicochemical interactions at the nano-bio interface. Nat. Mater. **8**(7), 543–557 (2009)
32. Kunath, K., et al.: Low-molecular-weight polyethylenimine as a non-viral vector for DNA delivery: comparison of physicochemical properties, transfection efficiency and in vivo distribution with high-molecular-weight polyethylenimine. J. Controlled Release **89**(1), 113–125 (2003)
33. Thomas, M., Klibanov, A.M.: Conjugation to gold nanoparticles enhances polyethylenimine's transfer of plasmid DNA into mammalian cells. Proc. Natl. Acad. Sci. **100**(16), 9138–9143 (2003)
34. Chasteen, N.D., Harrison, P.M.: Mineralization in ferritin: an efficient means of iron storage. J. Struct. Biol. **126**(3), 182–194 (1999)
35. Rice, R.H., Means, G.E.: Radioactive labeling of proteins in vitro. J. Biol. Chem. **246**(3), 831–832 (1971)

Chapter 5
Toxicological Profiling of Cationised Magnetoferritin

Toxicity of a drug or a material is a major concern in the context of any biomedical application. Nanoparticles have come under particular scrutiny, because toxic effects may become apparent when an otherwise non-toxic material is fabricated on the nanoscale [1]. In the context of regenerative medicine, exposure to harmful nanoparticles may significantly diminish the therapeutic efficiency of stem cell-based therapy. Therefore, careful toxicological evaluation is necessary. Magnetic nanoparticles can act on a variety of parameters that affect normal stem cell function, from interactions with the outer membrane of the cells to impacts on the DNA in the nucleus [2]. To assess toxicity, it is therefore important to perform a variety of toxicological tests that evaluate the impact of nanoparticle exposure on different cell functions [3, 4]. To date, the general consensus is that labelling stem cells with SPIONs is safe. This has been based on measurements of cell viability, proliferation and differentiation capacity, which have been shown to remain largely unaffected by SPION exposure (see Table 5.1). However, some MSC differentiation pathways seem to be more sensitive to SPION exposure than others, resulting in limited differentiation capacities [5, 6].

Although MF has previously been explored as a cellular contrast agent [12, 13], no detailed toxicological evaluation of this nanoparticle has been undertaken. In this chapter potential adverse effects of MF and cat-MF on a variety of hMSC functions are investigated. First, acute effects on cell viability and membrane integrity after cat-MF and MF exposure are presented. Acute toxic effects are defined as adverse outcomes of finite duration occurring up to 14 days after exposure to a single dose of a test substance [14]. Here, acute effects were investigated up to 48 h after exposure to MF and cat-MF. Next, results of a three week study investigating the impact of cat-MF exposure on cell proliferation and monolayer differentiation into adipocytes and osteoblasts are shown. Additionally, engineering of a cartilage tissue construct was undertaken using hMSCs labelled with cat-MF, as well as unlabelled hMSCs. Cartilage differentiation has been shown to be particularly affected by SPION exposure, even when other differentiation pathways were not [5]. Therefore, effects on cartilage differentiation following cat-MF exposure were

S. Correia Carreira, *Rapid Cell Magnetisation Using Cationised Magnetoferritin*,
Springer Theses, DOI 10.1007/978-3-319-60333-9_5

Table 5.1 A selection of previously reported toxicity outcomes in SPION-labelled MSCs

Particle	Size	Exposure dose	Result	Reference
Dextran coated iron oxide	80-150 nm	25 $\mu g\ mL^{-1}$, 100 $\mu g\ mL^{-1}$	No effect on cell viability after 12 h or 7 d	Kim et al. [7], Gasparetto et al. [8].
PLL[a]-coated Feridex[b]	Not stated	25 $\mu g\ mL^{-1}$	No effect on adipo- and osteogenesis, but chondorgenes is inhibited after a 24 h exposure	Kostura et al. [5].
PLL-coated iron oxide	15 nm	20 $\mu g\ mL^{-1}$	No effect on cell viability and proliferation after 1 and 24 h exposure	Ju et al. [9]
Citrate coated iron oxide	30 nm	2-10 mM	No effect on cell proliferation or membrane protein expression after a 2 h exposure	Wilhelm et al. [10]
Silica coated iron oxide	50 nm	30 $\mu g\ mL^{-1}$	No effect of cell proliferation after 1 h exposure	Lu et al. [11]

[a]PLL: Poly-(L-lysine)
[b]Feridex: commercially available dextran coated iron oxide

assessed qualitatively using histochemical staining and quantitatively using biochemical protein assays.

5.1 Acute Toxicity After MF and cat-MF Labelling

In this work, effects on cell viability 24 h after exposure to MF and cat-MF were investigated using the MTS and the Alamar Blue assay. Additionally, the Alamar Blue assay was also used to measure cell viability 4 and 48 h after exposure. Effects on cell membrane integrity were tested using an LDH assay conducted 24 h after exposure to MF and cat-MF. Details on the mechanism of these assays can be found in Sects. 2.12 and 2.13. Briefly, the MTS and the Alamar Blue assay give an indication of cell viability by measuring the formation of a coloured product as a result of electron transport reactions occurring during cell metabolism. The Alamar Blue assays offers the added benefit of repeated measurements on the same cell culture over time, which enables the identification of transient effects. The LDH assay gives an indication of cytotoxicity by measuring the amount of LDH enzyme in the cell culture supernatant. LDH is only present inside the cell and leaks out when the cell membrane integrity is compromised. The adequate cell numbers for the individual assays were determined prior to undertaking MF and cat-MF toxicity assessment, and results of the optimisation experiments can be found in Sects. 2.12.1 and 2.13.1.

5.1.1 Cell Viability and Cytotoxicity After MF and cat-MF Exposure

Effects on cell viability and membrane integrity were explored 24 h after exposure to 0.01 to 3 μM solutions of MF or cat-MF. The MTS assay, which uses cellular respiration as a measure of viability, showed no significant effects on cell viability after exposure to both MF and cat-MF across this concentration range ($F = 2.028$, $p = 0.068$), as well as no effect of surface functionalisation on the outcome of the MTS assay ($F = 0.079$, $p = 0.779$) (Fig. 5.1a). This was confirmed by an LDH assay, which showed that neither MF nor cat-MF led to cytotoxic effects affecting cell membrane integrity ($F = 0.951$, $p = 0.461$) and that surface functionalisation did not impact on the outcome of the assay ($F = 2.228$, $p = 0.138$) (Fig. 5.1b). The observation that similar viability levels were measured for MF and cat-MF indicates that cationisation did not confer additional cytotoxicity. This is a promising result, because cationic nanoparticles are often more cytotoxic than their anionic or neutral counterparts [15, 16]. This can be attributed to the relatively low zeta potential of cat-MF, which should avoid the hole formation observed in membranes exposed to nanoparticles with excessively high cationic charge density [17], e.g. with a zeta potential in excess of 20 mV [18].

The Alamar Blue assay was used to complement results of the MTS assay and extend viability monitoring to 4 and 48 h after exposure. Two-Way ANOVA indicated that surface functionalisation had an impact on how cell viability changed

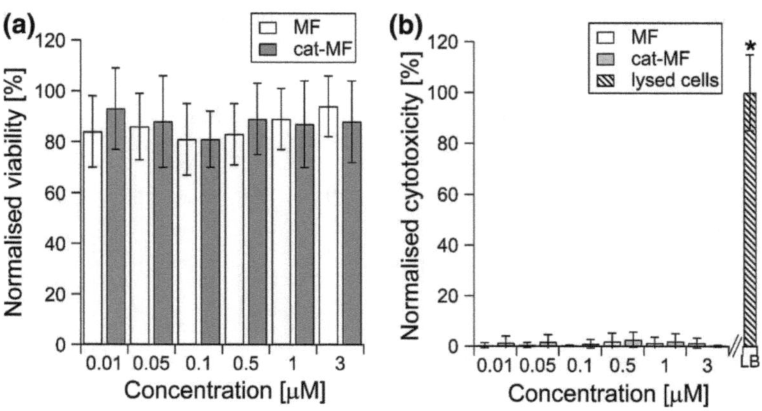

Fig. 5.1 Effects on hMSC viability and cytotoxicity 24 h after exposure to MF and cat-MF. **a** Viability and **b** cytotoxicity were assessed using an MTS and LDH assay, respectively, 24 h after exposure to MF and cat-MF for 30 min. Average and standard deviation of three biological replicates (i.e. hMSCs derived from three patients) tested in quadruplicate are shown (total n = 12 per group). All values were normalised to an untreated hMSC control. No significant effects of concentration or surface functionalisation on cell viability or cytotoxicity were found using Two-Way ANOVA. LB: lysis buffer used to disrupt the cell membrane and release intracellular LDH into the media (positive control). Image re-printed with permission from Correia Carreira et al. [19]

as a function of nanoparticle concentration 48 h ($F = 4.1$, $p = 0.001$), but not 4 h ($F = 1.474$, $p = 0.191$) or 24 h ($F = 1.153$, $p = 0.335$) after treatment. This was analysed in detail for each time point separately using a one-sided Dunnett t *post hoc* test. After 4 h, no reduction in cell viability was detected across the concentration range tested (Fig. 5.2a). However, 24 h after exposure to 3 μM MF and cat-MF, the test revealed a significant, 20% reduction in cell viability compared to untreated cells (Fig. 5.2b) (note that *post hoc* analysis also indicated that exposure to 0.5 μM MF had an effect, which is an improbable outcome seeing that 1 μM MF was measured to have no effect). After 48 h, cell viability remained reduced by 20% only in hMSCs that had been exposed to 3 μM cat-MF, whilst the MF-exposed hMSCs had recovered to 92% of the untreated control (Fig. 5.2c). This indicates that exposure to an elevated MF concentration had a transient effect, while a mild but significant effect of elevated cat-MF persisted over 48 h.

Taken together, these results indicate that both MF and cat-MF had only mild, if not no toxic effects on hMSCs 4, 24 and 48 h after exposure. These mild effects could be linked to elevated intracellular iron concentration. MF exposure led to transient effects, after which cell viability recovered to normal (unexposed) levels. This could be due to the fact that cells exposed to MF are likely to have internalised fewer nanoparticles compared to cell exposed to cat-MF (see Sect. 4.1), and therefore their total exposure burden was lower. A control experiment using ApoF and cat-ApoF could help elucidate whether this effect was actually mediated by the iron core or the cationic protein surface.

At the cat-MF concentration used for the previously presented hMSC magnetisation experiments (0.5 μM) no acute toxicity was observed. Additionally, the rapidity of cat-MF labelling means that short incubation times (less than 30 min)

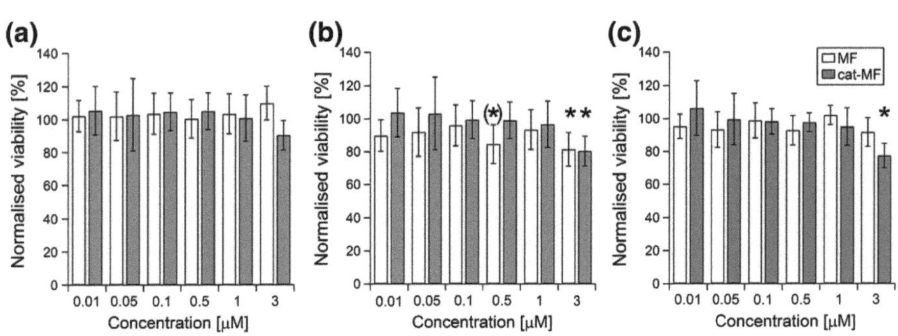

Fig. 5.2 Effects on cell viability 4, 24 and 48 h after exposure to MF and cat-MF. The Alamar Blue solution was added after a 30 min exposure, and subsequently fluorescence intensity was measured after **a** 4 h, **b** 24 h and **c** 48 h. Average and standard deviation of three biological replicates tested in quadruplicate are shown (total $n = 12$ per group). Values were normalised to an untreated hMSC control. Significant effects of exposure concentration were found using Two-Way Analysis of Variance and Dunnett t one-tailed *post hoc* test ($p < 0.05$), and indicated with an asterisk. Note: *post hoc* analysis indicated that exposure to 0.5 μM MF significantly reduced cell viability, however, this is an improbable outcome seeing that 1 μM MF was found to have no effect according to the same statistical analysis

can be used, which avoids cytotoxic effects associated with prolonged incubation time [20]. Furthermore, the nanoparticle concentration used in cat-MF labelling was several orders of magnitude lower than many competing SPION systems [21, 22], and the biocompatible protein shell of cat-MF is likely to circumvent cytotoxic effects that can arise from coating agents used to synthesise functionalised SPIONs [23].

The Alamar Blue assay proved to be more sensitive than the MTS assay, seeing that effects of high exposure concentrations were not detected using the latter method. Although both methods rely on very similar mechanisms, the detection of the assay product is different: in the MTS assay, the formation of a formazan product is monitored measuring absorbance, whereas in the Alamar Blue assay the formed resorufin molecule is detected using its fluorescence. Detecting a fluorescent signal can be more sensitive than absorbance measurements [24], which would explain the higher sensitivity of the Alamar Blue assay.

5.2 Long-Term Effects of cat-MF Labelling

The long-term effects of magnetising hMSCs with cat-MF were assessed over three and five weeks using a cell proliferation assay, as well as two differentiation assays. These parameters were tested because cell proliferation and differentiation capacity are two hallmarks of healthy MSCs [25].

5.2.1 Effects on Cell Proliferation

To evaluate the effect of cat-MF labelling on cell proliferation, the number of hMSCs grown in 75 cm [2] culture flasks after a 30 min exposure to 0.5 μM cat-MF or PBS (as an untreated control) was determined weekly up to three weeks. It was found that the cell number of both treated and untreated hMSCs increased to similar degrees from approximately 1.8 × 10 [5] to 1.1 × 10 [6] cells (Fig. 5.3a), and consequently no significant difference between the number of population doublings was found between treated and untreated groups (Fig. 5.3b) using the Wilcoxon Signed-Rank test ($p = 0.285$ for week 1 and 2, $p = 0.109$ for week 3). This result confirms that no toxicity is conferred on hMSCs exposed to 0.5 μM cat-MF.

5.2.2 Effects on Adipo- and Osteogenesis

Apart from the capacity for self-renewal, healthy stem cells have the capacity to differentiate into particular cell types (see Sect. 1.4). To assess whether

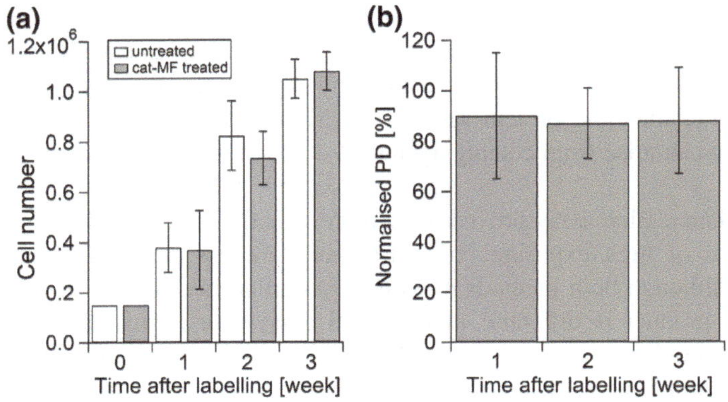

Fig. 5.3 Effect of cat-MF labelling on cell proliferation over three weeks. Proliferation of hMSCs was determined with a weekly cell counting assay after exposure to 0.5 μM cat-MF for 30 min, and found to be similar to proliferation of untreated cells. **a** Total number of cells counted each week. **b** Calculated population doublings (PD) normalised to the untreated hMSC control. Average and standard deviation of three biological replicates are shown

differentiation onto adipocytes (fat cells) and osteoblasts (bone cells) was affected, a monolayer differentiation assay was used. hMSCs were either exposed to PBS or 1 μM of cat-MF for 30 min, and then either cultured in regular medium or medium containing differentiation factors that induce formation of adipocytes or osteoblasts for three weeks. It was found that hMSCs treated with cat-MF differentiated into adipocytes and osteoblasts upon culture in differentiation medium (Fig. 5.4a, c), but

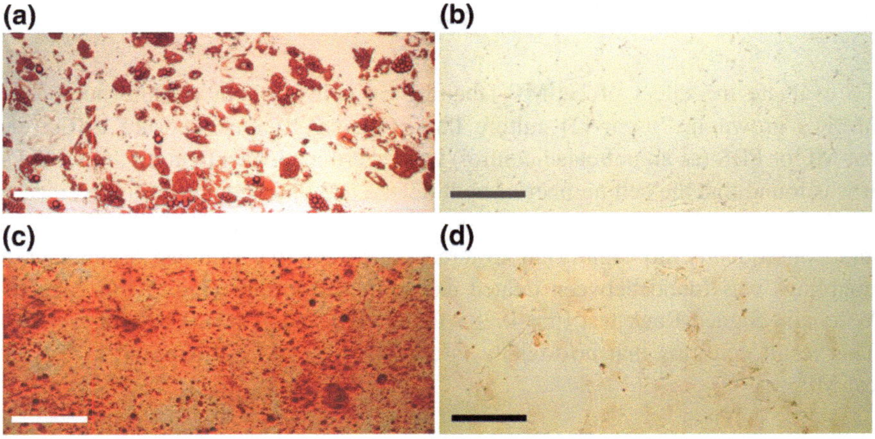

Fig. 5.4 Effect of cat-MF exposure on monolayer differentiation of hMSCs. Representative bright field microscopy images of **a** hMSC-derived adipocytes with fatty acid vacuoles stained using Oil Red. **b** Undifferentiated hMSCs stained with Oil Red. No fatty vacuoles are visible. **c** hMSC-derived osteoblasts with visible calcium phosphate deposits stained using Alizarin Red. **d** Undifferentiated hMSCs stained with Alizarin Red; no calcium phosphate deposits are visible. All *scale bars* 100 μm

remained undifferentiated when cultured in regular medium (Fig. 5.4b, d). The same was true for untreated hMSCs (see Appendix C).

This result is in line with previously reported effects of magnetic nanoparticles on hMSC differentiation into fat and bone cells [5]. However, the same study also found that cartilage differentiation (chondrogenesis) was inhibited after nanoparticle exposure. Therefore, in the final step of toxicological profiling it was investigated whether cartilage differentiation was affected after cat-MF exposure.

5.2.3 *Effects on Chondrogenesis*

Effects on cartilage formation were compared between untreated hMSCs and hMSCs exposed to 0.5 μM cat-MF for 30 min. Treated and untreated hMSCs were seeded onto fibronectin-coated PGA scaffolds and cultured for five weeks in medium containing the appropriate supplements to promote cartilage differentiation. Histochemical and immunohistochemical staining revealed the presence of proteoglycans (Fig. 5.5a) and collagen type II (Fig. 5.5b) in hMSCs labelled with cat-MF, as well as in the untreated control (Fig. 5.5c, d). (Note: unstained and H&E stained sections of both treatment groups can be found in Appendix D).

The development of cartilage is accompanied by the differentiation of MSCs into chondrocytes, which produce collagen type II and secrete proteoglycans. The expression of proteoglycans and collagen type II is therefore indicative of cartilage differentiation [26, 27].

To quantify the content of proteoglycans and collagen type II, biochemical analysis was performed on the engineered cartilage samples. Proteoglycans consist of a protein core, to which linear sulphated glycosaminoglycans (GAGs) are attached, most of them being chondroitin sulphate and keratin sulphate. The sulphated GAGs carry negative charges, which can interact with cationic dyes such as Safranin O or DMMB [28]. Therefore, GAG content in the engineered cartilage sample were quantified using a DMMB assay, whilst collagen type II was measured with an enzyme-linked immunosorbent assay (see Sects. 2.9.2 and 2.9.3). Cartilage samples were first freeze-dried and weighed, which revealed no significant difference in the dry mass of cartilage constructs engineered from untreated hMSCs compared to cartilage engineered from hMSCs treated with cat-MF (Mann-Whitney U test, $p = 1.0$; Fig. 5.6a). After that, cartilage constructs were enzymatically digested and the extracts were used to quantify GAG and collagen type II. Both were expressed as a percentage of the matrix dry weights and found to be similar in the two treatment groups (Fig. 5.6b; Mann-Whitney U Test, $p = 0.7$ for GAG, and $p = 1.0$ for collagen type II). GAG and collagen type II content was approximately 20 and 10% of the dry weight, respectively. These values are similar to previously reported GAG and collagen type II contents in engineered cartilage [29]. Taken together, these findings show that cat-MF exposure did not inhibit chondrogenesis.

Finally, cartilage sections were also stained with Prussian Blue to investigate whether cat-MF was still present as intact nanoparticles in the labelled cells after

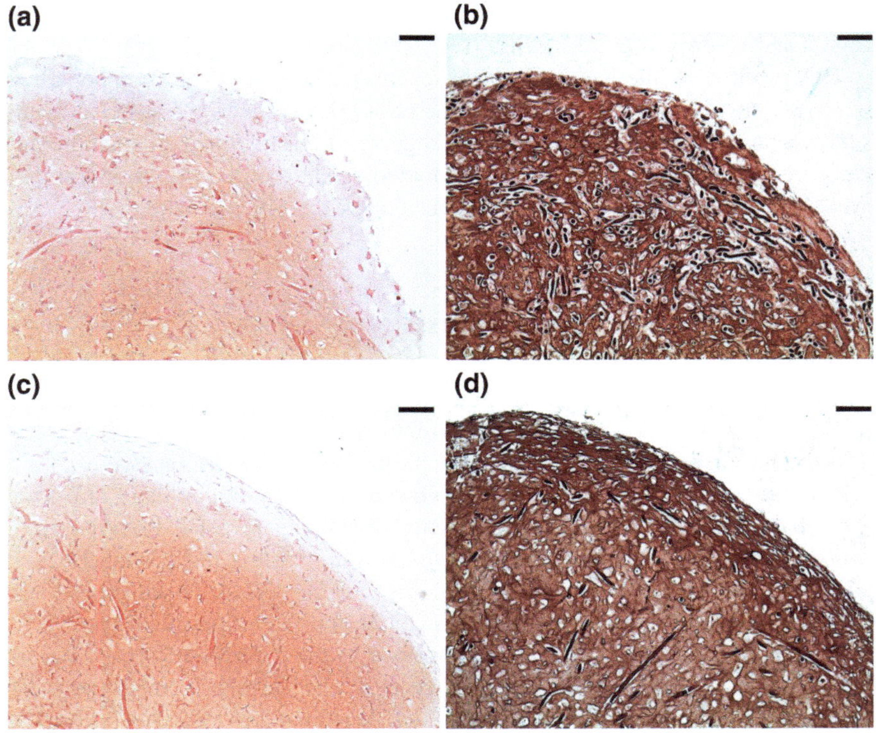

Fig. 5.5 Effect of cat-MF exposure on chondrogenesis. Representative bright field microscopy images of histological sections of engineered cartilage tissue constructs (**a**, **c**) stained for proteoglycans with Safranin O and (**b**, **d**) stained for collagen type II. *Panels A* and *B* show cartilage sections engineered from hMSCs exposed to 0.5 μM cat-MF for 30 min and *panels C* and *D* show cartilage sections grown from untreated hMSCs. All *scale bars*: 100 μm

five weeks in culture. Blue staining indicative of the presence of iron-rich nanoparticles was found in cartilage sections engineered from hMSC labelled with cat-MF (Fig. 5.7a), but not in cartilage grown from untreated hMSCs (Fig. 5.7c). Higher magnification imaging revealed that the blue stain was co-localised with the cell matrix and nuclei stained with Nuclear Fast Red (Fig. 5.7b), indicating that cat-MF could still be present in some cells. This is in agreement with the results reported in Sect. 4.3, which showed that 10% of hMSCs still contained sufficient amounts of cat-MF to be retained in a MACS column after five weeks in culture.

The finding that cat-MF labelling did not inhibit chondrogenesis is very significant, seeing that previous studies have shown that this differentiation pathway is affected by SPION exposure [5, 6]. In the study by Kostura et al. [5], inhibition of cartilage formation occurred in MSCs with similar cellular iron levels compared to the study presented here (approximately 13 pg per cell) [5]. In the study by Henning et al. [6], intracellular iron concentrations of approximately 5 pg per cell significantly reduced cartilage formation. Several parameters distinguish the work

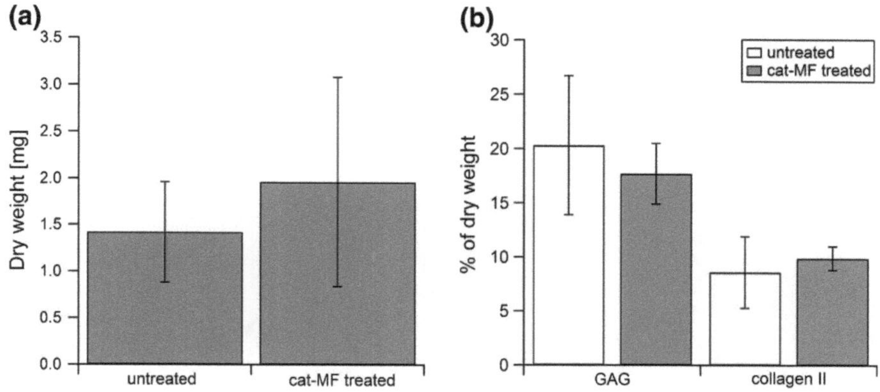

Fig. 5.6 Biochemical analysis of cartilage tissue constructs. **a** The dry weights of engineered cartilage tissue constructs derived from untreated hMSCs and hMCSc treated with 0.5 μM cat-MF. Average and standard deviation of three biological replicates are shown. **b** Quantification of glycosaminoglycans (GAG) and collagen II extracted from the dried cartilage tissue constructs. Average and standard deviation of three biological replicates are shown (with exception of the cat-MF treated collagen II value, which was determined from extracts of two cat-MF treated cartilage constructs due to insufficient sample volume to conduct the assay)

presented here from the exposure scenarios of the previous two studies: (1) The exposure concentration in this work was lower, namely 10 μg mL^{-} [1] iron, compared to 25 μg mL^{-} [1] used by Kostura et al. and 100 μg mL^{-} [1] used by Henning et al. [2] The exposure duration was shorter, namely 30 min in this work compared to 24 h used by Kostura et al. and 18 h used by Henning et al. [3]. The surface chemistry of the magnetic nanoparticle was different, cat-MF being encapsulated in a cationised protein coating, whereas Kostura et al. used cationic, poly-L-lysine functionalised SPIONs, and Henning et al. used anionic, carboxy-dextran functionalised SPIONs. It should be noted that Kostura et al. performed control experiments, in which MSCs were exposed to the poly-L-lysine coating only. This treatment did not inhibit chondrogenesis and thus the authors concluded that the iron core must have mediated toxicity. The authors also highlighted a study, which indicated that regulation of the iron metabolism is linked to chondrogenesis [30]. If iron homeostasis is important, and even 5 pg of iron per cell affect chondrogenesis, how is it possible that almost 15 pg of iron delivered to each cell inside cat-MF did not inhibit chondrogenesis? One possibility could be that the additional intracellular iron was not responsible for the inhibition but rather the prolonged exposure mediated the detrimental effects, e.g. by affecting cell signalling pathways. Alternatively, if the iron burden was to be considered the cause for inhibited chondrogenesis, the surface coatings may have played a role. In the work presented here, the magnetic nanoparticle was introduced in an iron storage protein, which may have been able to withstand acidic environments in the lysosome better than the coatings of the other two SPIONs. Thus, iron ions might have

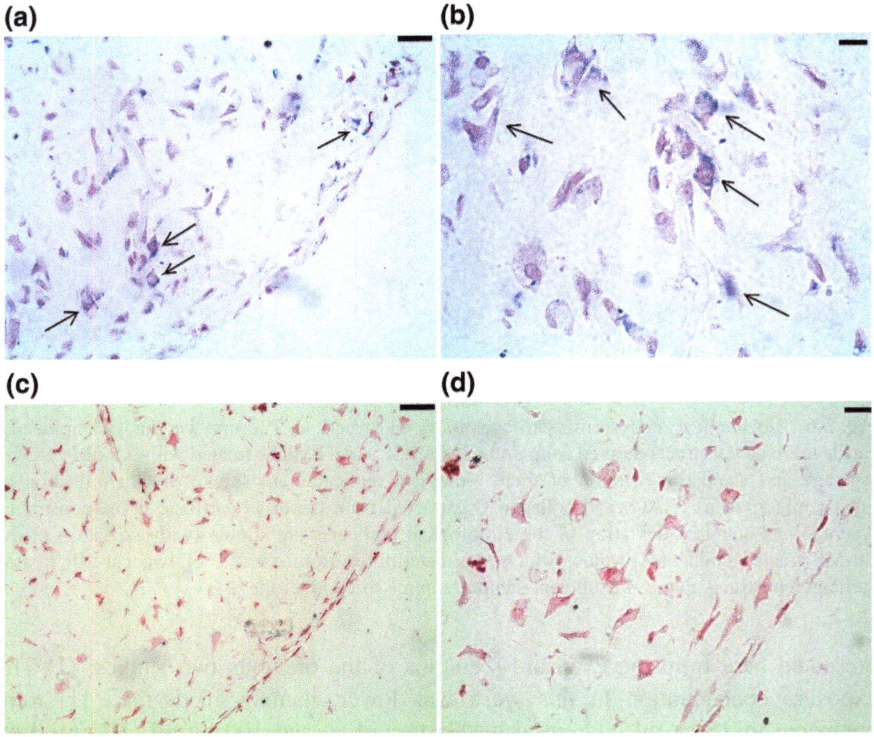

Fig. 5.7 Prussian Blue staining of cartilage tissue constructs. Representative bright field microscopy images of histological sections of cartilage constructs derived from (**a**, **b**) cat-MF labelled hMSCs and (**c**, **d**) untreated hMSCs. Sections were imaged using a (**a**, **c**) 20× objective, *scale bar* 50 μm and (B, D) 40× objective, *scale bar* 20 μm

been released more slowly, avoiding sudden disruption of the iron homeostasis. Either way, this result highlights that changes in exposure scenario dramatically influence toxicity outcomes.

5.3 Conclusions and Future Directions

Toxicological evaluation found that exposure to cat-MF at the concentrations used in the magnetic labelling experiments (see Chap. 4) had no detrimental effects on cell viability, membrane integrity, proliferation and differentiation of hMSCs. Significantly, chondrogenesis- a differentiation pathway that has previously been shown to be inhibited by exposure to magnetic nanoparticles- was not affected by cat-MF labelling.

Although cat-MF was found to have no significant adverse effects, there are a few points that remain unclear. For example, a mild reduction of cell viability was observed 24 h after exposure to elevated concentrations of cat-MF and MF. After 48 h, normal levels of cell viability had been restored in MF-exposed hMSCs, but not in cells exposed to cat-MF. It remained unclear whether this was due to differences in cellular iron content, or mediated by the cationised protein shell. A control experiment using ApoF and cat-ApoF could help elucidate this point.

It is also not clear why cat-MF exposure did not inhibit chondrogenesis although intracellular iron levels were as high as previously reported values that inhibited this differentiation route. One hypothesis is that the delivery of the magnetic nanoparticle in the ferritin cage may have altered the kinetics of iron release. This could be tested in a simple experiment, in which the lysosomal environment is simulated and the release of iron from cat-MF and dextran-coated SPIONs in monitored over time. A previous study has shown the feasibility of this approach to investigate lysosomal breakdown of SPIONs [31].

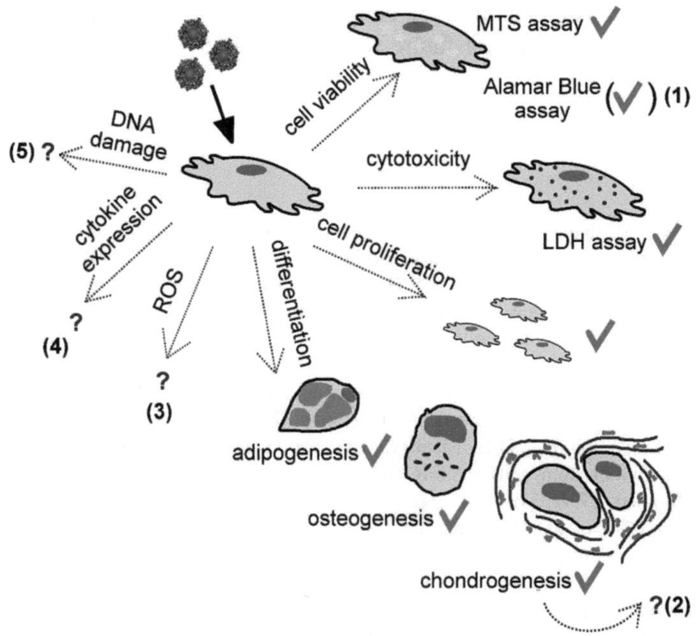

Fig. 5.8 Conclusions and directions for future work. Using the MTS assay, cell viability was found to be uncompromised after exposure to cat-MF. However, the Alamar Blue indicted a reduction in cell viability after exposure to elevated concentrations of cat-MF, but not MF. Control experiments are needed to elucidate whether this was mediated by the cationised surface or the iron core (*1*). Using the LDH assay, no cytotoxicity was detected after cat-MF exposure. Cell proliferation was also not compromised. The differentiation capacity of cat-MF-exposed hMSCs was unaffected, significantly, chondrogenesis was not inhibited. Whether this was a result of altered iron release kinetics, remains to be investigated (*2*). Further toxicological evaluation should examine if cat-MF exposure promotes the formation of reactive oxygen species (ROS) (*3*), impacts on cytokine expression (*4*) or leads to DNA damage (*5*)

Prussian Blue staining revealed that cat-MF particles could still be found in the engineered cartilage constructs five weeks after labelling. However, the Prussian Blue method relies on dissociation of iron ions in low pH, such that one cannot be absolutely sure that the Prussian Blue complex was formed due to the presence of nanoparticles or free iron ions. TEM imaging could reveal whether the cat-MF particles remained intact, although this method did not yield satisfactory results when hMSCs were observed one week after cat-MF labelling. Rather, it would be possible to measure the magnetic moment of the whole cartilage construct using SQUID magnetometry, which would give an indication of the number of intact magnetic nanoparticles present in the sample. A similar approach was taken to investigate long-term SPION degradation in tissues of a mouse model [32].

Further toxicological evaluation of MF and cat-MF should investigate the potential to promote the production of reactive oxygen species, which is a common pathway of metal and metal oxide nanoparticle toxicity that can adversely affect many cell functions [33]. Furthermore, the impact on cytokine expression and DNA damage should also be explored, the latter being of particular importance regarding potential detrimental outcomes in daughter cells [34]. A graphic summary of conclusions and directions for future work can be found in Fig. 5.8.

References

1. Jiang, W., Mashayekhi, H., Xing, B.: Bacterial toxicity comparison between nano-and micro-scaled oxide particles. Environ. Pollut. **157**(5), 1619–1625 (2009)
2. Markides, H., Rotherham, M., El Haj, A.: Biocompatibility and toxicity of magnetic nanoparticles in regenerative medicine. J. Nanomater. **2012**, 13 (2012)
3. Dusinska, M., et al.: Towards an alternative testing strategy for nanomaterials used in nanomedicine: lessons from nanotest. Nanotoxicology **9**(supl), 118–132 (2015)
4. Kim, J.-E., Shin, J.-Y., Cho, M.-H.: Magnetic nanoparticles: an update of application for drug delivery and possible toxic effects. Arch. Toxicol. **86**(5), 685–700 (2012)
5. Kostura, L., Kraitchman, D.L., Mackay, A.M., Pittenger, M.F., Bulte, J.W.: Feridex Labeling of mesenchymal stem cells inhibits chondrogenesis but not adipogenesis or osteogenesis. NMR Biomed. **17**(7), 513–517 (2004)
6. Henning, T.D., et al.: The influence of ferucarbotran on the chondrogenesis of human mesenchymal stem cells. Contrast Media Mol. Imaging **4**(4), 165–173 (2009)
7. Kim, T.H., Kim, J.K., Shim, W., Kim, S.Y., Park, T.J., Jung, J.Y.: Tracking of transplanted mesenchymal stem cells labeled with fluorescent magnetic nanoparticle in liver cirrhosis rat model with 3-T MRI. Magn. Reson. Imaging **28**(7), 1004–1013 (2010)
8. Gasparetto, E.L., Spray, D.C., de Carvalho, A.C.C., Mendez-Otero, R.: Optimized labeling of bone marrow mesenchymal cells with superparamagnetic iron oxide nanoparticles and in vivo visualization by Magnetic Resonance Imaging (2011)
9. Ju, S., Teng, G., Zhang, Y., Ma, M., Chen, F., Ni, Y.: In vitro labeling and mri of mesenchymal stem cells from human umbilical cord blood. Magn. Reson. Imaging **24**(5), 611–617 (2006)
10. Wilhelm, C., et al.: Magnetic control of vascular network formation with magnetically labeled endothelial progenitor cells. Biomaterials **28**(26), 3797–3806 (2007)
11. Lu, C.-W., et al.: Bifunctional magnetic silica nanoparticles for highly efficient human stem cell labeling. Nano Lett. **7**(1), 149–154 (2007)

12. Uchida, M., et al.: A human ferritin iron oxide nano-composite magnetic resonance contrast agent. Magn. Reson. Med. **60**(5), 1073–1081 (2008)
13. Bulte, J.W.M., et al.: Magnetoferritin—characterization of a novel superparamagnetic MR contrast agent. JMRI: J. Magn. Reson. Imaging **4**(3), 497–505 (1994)
14. McNaught, A.D., McNaught, A.: Compendium of Chemical Terminology, vol. 1669. Blackwell Science Oxford (1997)
15. Goodman, C.M., McCusker, C.D., Yilmaz, T., Rotello, V.M.: Toxicity of gold nanoparticles functionalized with cationic and anionic side chains. Bioconjug. Chem. **15**(4), 897–900 (2004)
16. Frohlich, E.: The role of surface charge in cellular uptake and cytotoxicity of medical nanoparticles. Int. J. Nanomed. **7**, 5577–5591 (2012)
17. Nel, A.E., et al.: Understanding biophysicochemical interactions at the nano-bio interface. Nat. Mater. **8**(7), 543–557 (2009)
18. Shao, X.R., et al.: Independent effect of polymeric nanoparticle zeta potential/surface charge, on their cytotoxicity and affinity to cells. Cell Prolif. **48**(4), 465–474 (2015)
19. Carreira, S.C., Armstrong, J., Seddon, A., Perriman, A., Hartley-Davies, R., Schwarzacher, W.: Ultra-fast stem cell labelling using cationised magnetoferritin. Nanoscale **8**(14), 7474–7483 (2016)
20. Singh, N., Jenkins, G.J., Asadi, R., Doak, S.H.: Potential toxicity of superparamagnetic iron oxide nanoparticles (Spion). Nano Rev. **1** (2010)
21. Liong, M., et al.: Multifunctional inorganic nanoparticles for imaging, targeting, and drug delivery. ACS Nano **2**(5), 889–896 (2008)
22. Wilhelm, C., Billotey, C., Roger, J., Pons, J., Bacri, J.-C., Gazeau, F.: Intracellular uptake of anionic superparamagnetic nanoparticles as a function of their surface coating. Biomaterials **24**(6), 1001–1011 (2003)
23. Häfeli, U.O., et al.: Cell uptake and in vitro toxicity of magnetic nanoparticles suitable for drug delivery. Mol. Pharm. **6**(5), 1417–1428 (2009)
24. Gallagher, S.R., Desjardins, P.R.: Quantitation of DNA and RNA with absorption and fluorescence spectroscopy. Curr. Protocols Hum. Genet. A. 3D. 1-A. 3D. 21 (2006)
25. Young, H.E., Black, A.C.: Adult stem cells. Anat. Rec. A Discov. Mol. Cell. Evol. Biol. **276A**(1), 75–102 (2004)
26. Pittenger, M.F., et al.: Multilineage potential of adult human mesenchymal stem cells. Science **284**(5411), 143–147 (1999)
27. Schmitz, N., Laverty, S., Kraus, V., Aigner, T.: Basic methods in histopathology of joint tissues. Osteoarthritis Crtilage **18**, S113–S116 (2010)
28. Roughley, P.J., Lee, E.R.: Cartilage proteoglycans: structure and potential functions. Microsc. Res. Tech. **28**(5), 385–397 (1994)
29. Dunkelman, N.S., Zimber, M.P., LeBaron, R.G., Pavelec, R., Kwan, M., Purchio, A.: Cartilage production by rabbit articular chondrocytes on polyglycolic acid scaffolds in a closed bioreactor system. Biotechnol. Bioeng. **46**(4), 299–305 (1995)
30. Nakamasu, K., et al.: Membrane-bound transferrin-like protein (Mtf): structure, evolution and selective expression during chondrogenic differentiation of mouse embryonic cells. Biochimica et Biophysica Acta (BBA)-Gene Struct. Exp. **1447**(2), 258–264 (1999)
31. Arbab, A.S., Wilson, L.B., Ashari, P., Jordan, E.K., Lewis, B.K., Frank, J.A.: A model of lysosomal metabolism of dextran coated superparamagnetic iron oxide (Spio) nanoparticles: Implications for cellular magnetic resonance imaging. NMR Biomed. **18**(6), 383–389 (2005)
32. Kolosnjaj-Tabi, J., et al.: The one year fate of iron oxide coated gold nanoparticles in mice. ACS Nano (2015)
33. Naqvi, S., et al.: Concentration-dependent toxicity of iron oxide nanoparticles mediated by increased oxidative stress. Int. J. Nanomed. **5**, 983 (2010)
34. Singh, N., et al.: Nanogenotoxicology: The DNA damaging potential of engineered nanomaterials. Biomaterials **30**(23), 3891–3914 (2009)

Chapter 6
Escherichia coli Labelling with Cationised Magnetoferritin

Infectious diseases are among the world's most pressing health challenges [1], and developing strategies for fast identification of pathogens is particularly important for effective treatment. However, current methodologies rely on culture-based microbiological methods that take several days. Rapid identification of pathogens can be achieved based on the detection of specific genes [2] or by antibody binding of pathogens on a sensor surface [3]. Microfluidic devices have shown to be excellent platforms for cheap and high-throughput implementation of these methods [3, 4]. Microfluidic devices typically handle volumes of 50–200 μL, but clinical blood or urine samples usually range between 5 and 20 mL in volume. Therefore, concentration of pathogens from large into small volumes represents an important processing step in microfluidics- based diagnostics. Magnetic labelling of bacteria in suspension enables their extraction from aqueous samples using a magnetic field, such that low numbers of bacteria can be concentrated [5].

In this chapter, the potential of cat-MF to magnetise *Escherichia coli* (*E. coli*) cells in water samples and immobilising them in MACS columns is explored. The use of MACS columns has traditionally been restricted to the capture of magnetically labelled mammalian cells, and extension of this method to the capture of bacteria has not yet been reported. *E. coli* has been identified as the most adequate indicator for drinking water quality [6], it is therefore important to detect even low numbers of these bacteria in water samples. *E. coli* is a gram negative bacterium and its outer membrane contains lipopolysaccharides, which impart a negative surface charge to the bacterium [7]. Therefore, electrostatic interactions between the cat-MF surface and the anionic bacterial cell wall should facilitate rapid magnetisation, as observed in hMSCs (see Chap. 4). First, results on the rapidity and concentration dependence of magnetic labelling with cat-MF using a MACS column are presented. Subsequently, capture efficiency of the MACS column is explored using various concentrations of cat-MF labelled *E. coli*, as well as the potential to magnetically concentrate low numbers of *E. coli*. In addition, magnetic labelling of *Staphylococcus aureus* with cat-MF and subsequent immobilisation in MACS columns is compared to the capture efficiency of magnetised *E. coli*.

S. Correia Carreira, *Rapid Cell Magnetisation Using Cationised Magnetoferritin*,
Springer Theses, DOI 10.1007/978-3-319-60333-9_6

Finally, results on the effect of exposure to a variety of cat-MF concentrations on *E. coli* growth are presented.

6.1 Time and Concentration-Dependence of Magnetising *E. coli* with cat-MF

Magnetic capture was investigated by incubating 0.6 mL water samples inoculated with *E. coli* for 1 to 30 min with 0.5 μM cat-MF and passing the water samples through a MACS column. Samples from the initial bacterial suspension (contaminated water) and from the flow-through of the column (decontaminated water) were collected and absorbance was measured at 600 nm. 100 μL of these samples were also plated out in triplicate onto LB agar to assess the number of colony forming units still present in the decontaminated water and to calculate the percentage of bacteria removed by magnetic separation. The absorbance of the decontaminated water samples shows a dramatic decrease compared to the absorbance of the contaminated sample after an incubation period of only 1 min (Fig. 6.1a). Similarly, it was found that 99.968 ± 0.006% of E. coli were removed from the water sample by magnetic capture after a 1 min incubation with cat-MF (Fig. 6.1b). After 30 min, 99.995 ± 0.001% were removed.

This result represents an improvement in bacterial capture efficiency compared to conventional setups that consist of a permanent magnet placed against a vial. In these setups, the magnetically labelled bacteria agglomerate in the applied magnetic

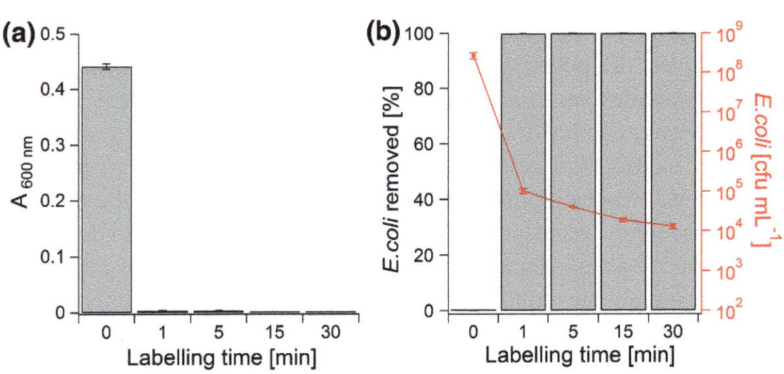

Fig. 6.1 Magnetic capture of *E. coli* after different incubation intervals with cat-MF. **a** Absorbance at 600 nm of water samples inoculated with *E. coli* and incubated with 0.5 μM cat-MF for 1 to 30 min. Average and standard deviation of three UV/Vis measurements are shown. **b** Percentage removal of *E. coli* from the contaminated water sample and number of *E. coli* still present in the water after MACS determined with the plate count method. The "0" time point shows values measured in the contaminated water sample prior to magnetic separation, whereas the time points "1" to "30" represent values measured in water samples passed through the MACS column at various time points after incubation with cat-MF. Average and standard deviation of three plate counts are shown

field, such that the supernatant can be removed and analysed for any residual bacteria. In Table 6.1 results from studies using this setup are compared to the results presented here using MACS. The improvement of capture efficiency may be due to the generation of strong local magnetic fields by the steel beads in the MACS column. In a setup that consists of a permanent magnet placed against a vial, the magnetic field decays rapidly with the distance from the magnet. Therefore, the magnetic field may not be strong enough to agglomerate magnetised bacteria from the entire volume. The advantage of this simpler setup over the use of MACS columns is that large volumes can be handled in relatively short amounts of time. Processing of larger volumes in the MACS columns is restricted by the flow rate of about 0.5 mL min^{-1}. However, MACS column could be scaled up, such that larger samples may also be processed.

It must be noted that removal of *E. coli* from inoculated water samples was not complete, approximately 10^4 cfu mL^{-1} were still present in the water sample after MACS. This is a relatively high bacterial concentration and water contaminated to such a degree would not be fit for drinking. Faecal indicator bacteria such as *E. coli* are present at high concentrations in wastewater (10^6–10^{10} organisms per litre), and using *E. coli* as an indicator for faecal contamination is common practice in the assessment of drinking-water quality [11]. Water intended for drinking should contain no faecal indicator organisms. Furthermore, water-borne infections by pathogens such as *E. coli* O157 and other enterohaemorrhagic strains of *E. coli* can be life threatening. The infective dose is as low as 100 organisms [11], highlighting the need for complete removal of faecal indicator bacteria from water intended for human consumption.

TEM imaging of *E. coli* eluted from the magnetic column after 1 min incubation with cat-MF revealed the presence of electron-dense nanoparticles of approximately 5 to 6 nm diameter on the surface of the bacterium (Fig. 6.2a). It was concluded that these particles were cat-MF, because no nanoparticles were visible in untreated *E. coli* samples (Fig. 6.2b).

Table 6.1 Comparison of presented *E. coli* capture efficiency with previously reported results

SPION surface	Exposure dose	Exposure time	Capture efficiency	*E. coli* detection	References
Amine	10 µg mL^{-1}	1 min	99.97%	Plate counting	This work
Amine	1 mg mL^{-1}	1 min	97%	A_{600nm}^{a}	Huang et al. [8]
Carboxyl	2 mg mL^{-1}	12 h	35%	A_{600nm}	Singh et al. [9]
Mannose	2 mg mL^{-1}	45 min	88%	Microscopy	Boubbou et al. [10]
Vancomycin	0.2 µg mL^{-1}	60 min	83%	Plate counting	Kell et al. [5]

[a] A_{600nm}: absorbance measured at 600 nm

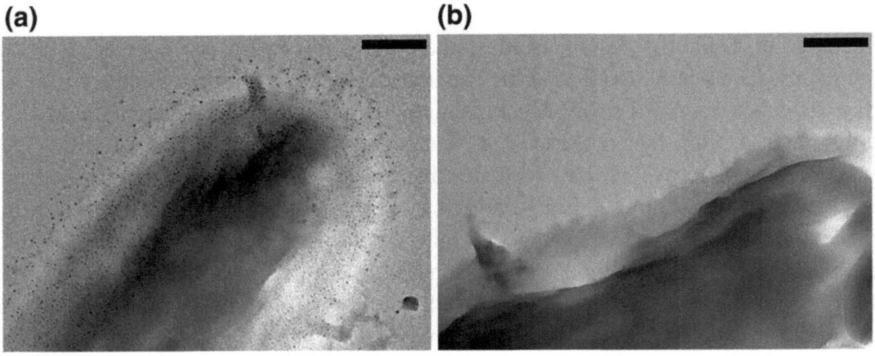

Fig. 6.2 TEM of *E. coli*. **a** *E. coli* exposed to 0.5 μM cat-MF for 1 min, immobilised and then eluted from the MACS column. Electron-dense nanoparticles with a diameter matching cat-MF uniformly cover the surface of the bacterium. **b** Untreated *E. coli*. No nanoparticles are visible on the surface. *Scale bars* 200 nm. Image re-printed with permission from Correia Carreira et al. [12]

Cationised ferritin has been used as a TEM "stain" in a variety of studies investigating the distribution of anionic domains on the surface of bacteria, including *E. coli* [13, 14]. Similarly, electrostatic interactions are likely to have mediated the adsorption of cat-MF onto anionic domains on the bacterial surface as observed here. These domains could be negatively charged polysaccharides in the lipopolysaccharide layer of *E. coli,* such as hexuronic acid and other acidic sugars [14, 15].

Next, the concentration dependent decontamination efficiency was investigated by incubating water samples inoculated with *E. coli* with cat-MF concentrations ranging from 0.01 to 1 μM for 15 min. The surface area of *E. coli* is 1.6×10^{-4} mm [2, 7] therefore, it can bind approximately 10^7 cat-MF particles. The number of cat-MF particles per cell required to overcome the drag force in the column under flow is 7.8×10^4 (see Sects. 2.11.2 and 4.1), therefore, the bacterial surface has ample space to accommodate the number of cat-MF particles required to magnetise the cell sufficiently for immobilisation in the MACS column. The maximum removal efficiency was 99.99%, which was achieved using 0.1 μM cat-MF (Fig. 6.3a). Higher cat-MF concentrations did not improve capture efficiency, which could be due to saturation of the MACS column with bacteria (see Sect. 6.2 for a further investigation of this hypothesis). Interestingly, cat-MF concentrations as low as 0.01 μM resulted in the capture of 98% of *E. coli*. Given that each *E. coli* cell only needs to bind 7.8×10^4 particles to be magnetised sufficiently, and that 4×10^8 cells are present in the inoculated sample, a total of 3.1×10^{13} particles would be needed label each cell sufficiently. A solution of 0.01 μM contains 6×10^{15} cat-MF particles, which should be sufficient to achieve labelling of all bacteria. However, TEM images of *E. coli* exposed to 0.01 μM cat-MF reveal that the distribution of nanoparticles on the cell surface appears patchy (Fig. 6.3b), whilst incubation with 0.1 μM cat-MF resulted in a relatively even distribution of cat-MF all over the cell surface (Fig. 6.3c).

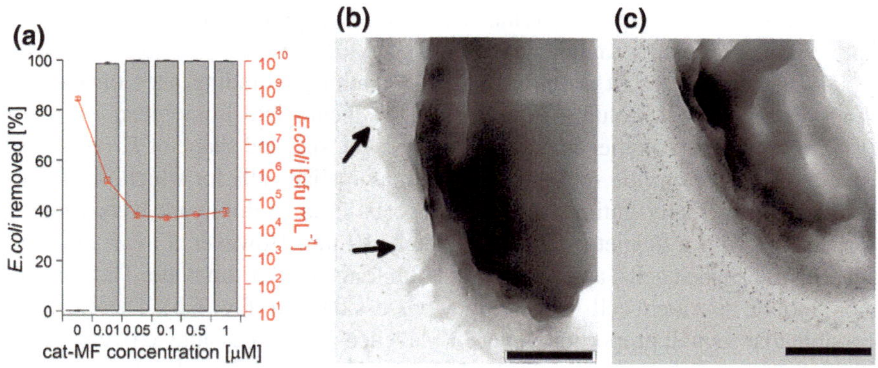

Fig. 6.3 Magnetic labelling of *E. coli* using a range of cat-MF concentrations. **a** Percentage removal of *E. coli* from the contaminated water sample (*grey bars*) and number of *E. coli* still present in the water after MACS (*red lines* and markers). Average and standard deviation of three plate counts are shown. **b** TEM image of *E. coli* exposed to 0.01 μM cat-MF. *Arrows* point to patches of nanoparticles on the cell surface. **c** TEM image of *E. coli* exposed to 0.1 μM cat-MF. Nanoparticles are distributed evenly over the cell surface. *Scale bars* 200 nm

Similar observations have been made previously, for example incubating mycoplasma cells with low concentrations of cationised ferritin (35 μg per mg of mycoplasma protein) showed irregular binding on small regions of the cell surface, whereas incubation with higher concentrations (140 μg per mg of mycoplasma protein) resulted in a uniform distribution of the particles on the cell surface. [16] It has been suggested that this behaviour is a result of affinity differences of the various anionic sites present on the bacterial surface. Regions with high affinity for cations are more effective at binding low concentrations of cationic particles, and thus, cationic affinity distributions on the cell surface could be assessed using decreasing amounts of cationised ferritin [13]. This "patchy" affinity binding at low cat-MF concentrations may explain the lower capture efficiency.

6.2 Magnetic Capture of Various Concentrations of *E. coli*

In Fig. 6.3 it was shown that cat-MF could be used to magnetise *E. coli* such that over 99.9% of the bacteria could be removed from the water sample by passing it through a MACS column. However, 100% capture was not achieved, even when incubating *E. coli* with higher concentrations of cat-MF. Therefore, it was hypothesised that the MACS column may have been saturated with bacteria. To test this hypothesis, removal efficiency was investigated as a function of bacterial contamination. Water samples were inoculated with 1×10^2 to 2×10^8 cfu mL^{-1} *E. coli*, incubated with 0.5 μM cat-MF for 15 min, and passed through a MACS column. It was observed that the number of *E. coli* removed from the contaminated water sample was proportional to the amount of bacteria present in the sample

(Fig. 6.4a). Removal efficiency remained constant at approximately 99.9% and was only complete in water samples inoculated with 10^2 cfu mL^{-1} *E. coli* (Fig. 6.4b).

Given these results, it is unlikely that saturation of binding sites on the steel beads of the MACS column is the cause for incomplete immobilisation of the *E. coli* population. Another factor to consider is the size and shape of *E. coli*, which is very different from the size and shape of mammalian cells for which the column design was originally optimised. The stem cells used in this work were round in suspension, with a diameter of approximately 10 μm. However, *E. coli* are much smaller, rod shaped cells, approximately 1 μm long and 0.5 μm wide. They are also more mobile than stem cells because they possess flagella (Fig. 6.5). Therefore, it is possible that a small proportion of cat-MF labelled *E. coli* was able to overcome the magnetic force acting on them in the MACS column (see Sect. 6.3 for an experiment using a non-flagellated bacterium). Several samples of the flow-through from the MACS columns were investigated in TEM to verify whether the residual *E. coli* had bound any cat-MF on their surface. However, no bacteria could be found, presumably because the concentration of *E. coli* was very low. Thus, the reason for incomplete capture of the magnetised bacterial population remain unclear.

Next, it was investigated if low bacterial numbers (under 10^2 cfu mL^{-1}) could still be completely removed from a water sample and concentrated using immobilisation in the MACS column and elution into a small volume. Detection of bacteria at concentrations of less than 10^2 cfu mL^{-1} is notoriously difficult without pre-enriching bacteria through a culture process [17]. Therefore, capture of low amounts of bacteria and their immediate concentration represents a significant step towards rapid detection and identification of pathogens. A 2 mL water sample

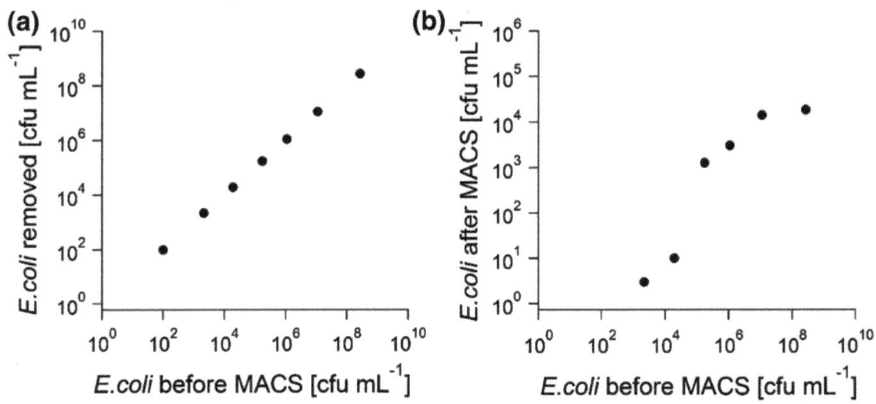

Fig. 6.4 Magnetic capture of various numbers of *E. coli* from water samples. Water samples were inoculated with 1×10^2 to 2×10^8 cfu mL^{-1} *E. coli* and subsequently incubated with 0.5 μM cat-MF for 15 min. **a** The number of *E. coli* removed by MACS as a function of bacteria initially present in the water sample before MACS. **b** The number of residual *E. coli* still present in water samples after MACS as a function of bacteria initially present in the water before MACS. Removal efficiency remained constant at approximately 99.9% and was only complete in water samples inoculated with 10^2 cfu mL^{-1} *E. coli*

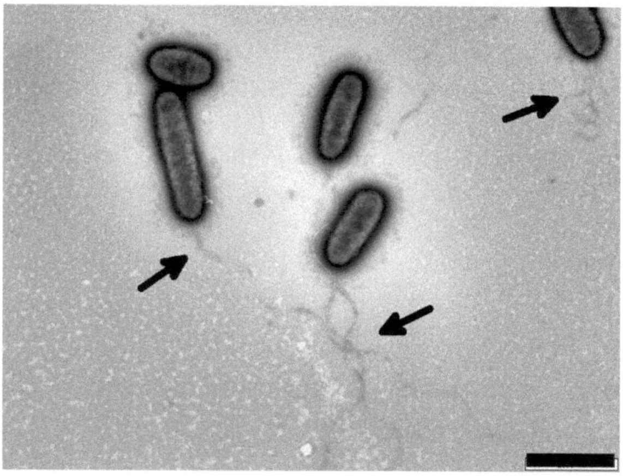

Fig. 6.5 TEM of *E. coli* stained with 2% phosphotungstic acid. *Arrows* indicate the flagella of the bacteria. *Scale bar* 2 μm

containing approximately 50 cfu mL^{-1} *E. coli* was incubated with 0.5 μM cat-MF for 15 min and passed through a MACS column. No bacteria could be detected in the flow-through after MACS (Fig. 6.6). When the column was removed from the magnet and the immobilised bacteria eluted using 0.4 mL sterile water, approximately 320 cfu mL^{-1} *E. coli* were detected, which represents an almost sevenfold concentration increase of the bacteria.

Previously, SPIONs have been functionalised with vancomycin to bind to the surface of a variety of bacteria. Vancomycin is a glycopeptide molecule that interacts strongly with the cell surface of gram-positive bacteria through hydrogen

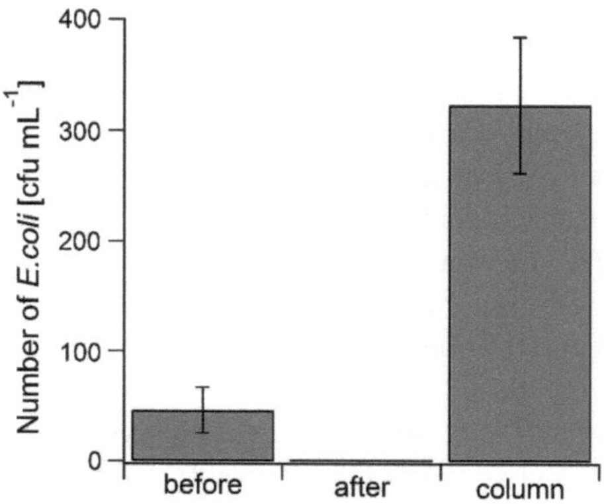

Fig. 6.6 Removal and concentration of dilute *E. coli*. A 2 mL water sample containing 46 cfu mL^{-1} *E. coli* (labelled "before") was incubated with 0.5 μM cat-MF for 15 min. After MACS, no bacteria were detected in the flow-through (labelled "after"). *E. coli* immobilised in the MACS column were eluted with 0.4 mL sterile water, through which a sevenfold concentration increase of bacteria was achieved (labelled "column"). Average and standard deviation from three plate counts are shown. Image re-printed with permission from Correia Carreira et al. [12]

bonding but has also been found to bind to gram-negative bacteria, presumably due to unspecific binding [18]. It has to be noted that vancomycin conjugation onto SPIONs is not trivial, and capture efficiency depends strongly on the orientation of the molecule on the SPION [5]. Using vancomycin-functionalised SPIONs, the lowest concentration of *E. coli* that could be captured was 10 cfu mL^{-1}. This is of the same order of magnitude as the lowest concentration of *E. coli* captured here ($\sim$50 cfu mL^{-1}), which was achieved using a much more facile surface functionalisation.

6.3 Comparing Magnetic Capture of *E. coli* and *S. aureus* Using cat-MF

The capacity for cat-MF to label other bacteria for capture in MACS columns was explored using *S. aureus*. *S. aureus* is a gram-positive bacterium, which is a class of bacteria that lack the outer lipopolysaccharide layer that is characteristic for gram negative bacteria. Thus, they possess a thick outer peptidoglycan layer to stabilise their shape and protect them from the environment [19]. *S. aureus* has been shown to have a negative zeta potential [20], therefore, negative charges should be available for electrostatic binding of cat-MF to domains on the peptidoglycan surface.

Water samples inoculated with approximately 10^8 cfu mL^{-1} *E. coli* and *S. aureus* were incubated with 0.5 μM cat-MF for 15 min and passed through MACS columns. Capture efficiency was higher for *S. aureus* than for *E. coli* (Fig. 6.7). It was found that 99.99% of cat-MF labelled *E. coli* could be captured using MACS, compared to 99.9999% of cat-MF labelled *S. aureus*. This is a significant finding given that it was previously claimed that magnetic capture of *S. aureus* was not possible using amine-functionalised (*i.e.* cationised) SPIONs and that small molecule probes were required to achieve this [21].

There are a number of characteristics that distinguish *S. aureus* from *E. coli* and that could have influenced capture efficiency. Firstly, *S. aureus* is a gram positive bacterium and therefore the make-up of its cell surface is different [19]. Secondly, the surface area of *E. coli* is approximately 163×10^{-6} mm^2 whereas *S. aureus* has a surface area of 33×10^{-6} mm^2 [7]. Thirdly, *E. coli* is a rod-shaped bacterium while *S. aureus* is spherical (Fig. 6.8). Finally, the *E. coli* strain used here is flagellated (Fig. 6.5) and therefore mobile in aqueous environments, but *S. aureus* has no such structure and is therefore less mobile. Considering the cell surface, it has been shown that both *E. coli* and *S. aureus* possess a very similar zeta potential [20], so the affinity for binding cat-MF may be similar in both species. However, *E. coli* has a much greater surface area, therefore more cat-MF particles are able to bind, imparting a larger magnetisation to the cell. Given that fewer *E. coli* could be immobilised in the MACS column compared to *S. aureus*, this potentially larger magnetisation didn't seem to play a role in improving capture efficiency.

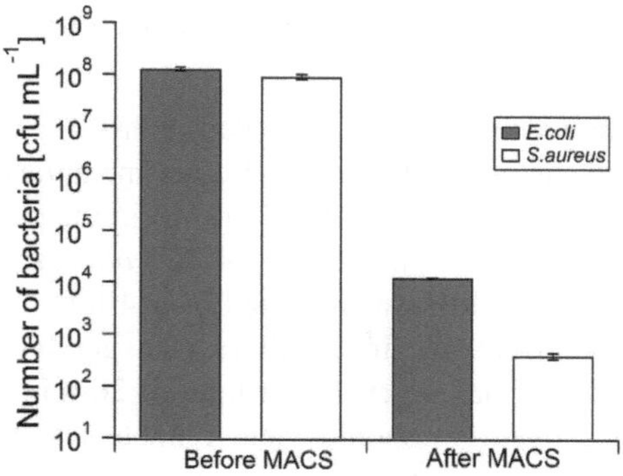

Fig. 6.7 Magnetic capture of *E. coli* and *S. aureus* labelled with cat-MF. Water samples were inoculated with approximately 10^8 cfu mL^{-1} and incubated with 0.5 μM cat-MF for 15 min before MACS. After MACS, approximately 10^4 cfu mL^{-1} of *E. coli* were still present in the sample after MACS, compared to only 4×10^2 cfu mL^{-1} of *S. aureus*. Average and standard deviation from three plate counts are shown

Considering the size and shape of both species, *S. aureus* has a spherical shape and a diameter of 0.6 μm, whilst *E. coli* is rod-shaped, approximately 1 μm long and 0.5 μm wide [22]. This may influence the drag force acting on the two species, with the smaller *S. aureus* experiencing less drag. Finally and possibly most importantly, *E. coli* has the ability to actively move using its flagella whereas *S. aureus* has no such means of autonomous locomotion. It may therefore be possible for some *E. coli* to escape the magnetic field, whereas *S. aureus* remain immobilised.

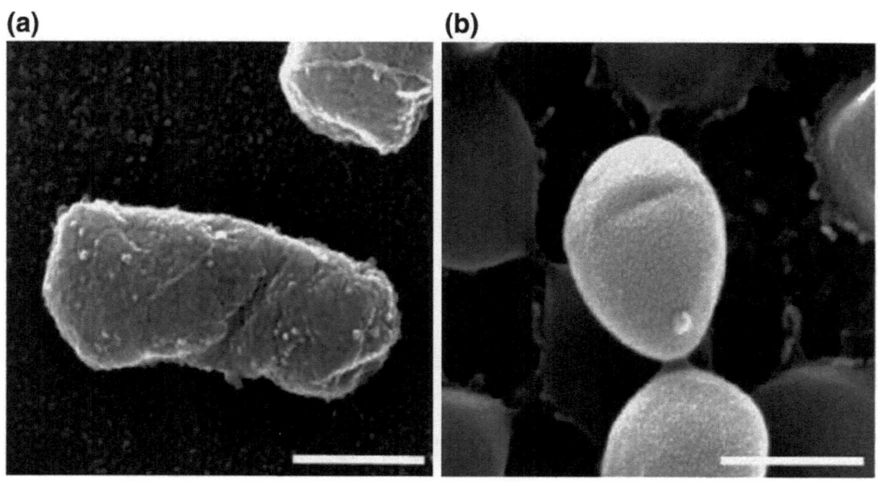

Fig. 6.8 Scanning electron micrograph of (A) *E. coli* and (B) *S. aureus*. *Scale bar* 500 nm. Image adapted with permission from Otto et al. [23]

6.4 The Effect of cat-MF Exposure on *E. coli* Growth

In Fig. 6.3 it was shown that *E. coli* incubated with 0.1 μM or more of cat-MF resulted in the even coverage of the bacterial outer membrane with the nanoparticles. To investigate whether the presence of cat-MF in the outer membrane had an effect on bacterial growth, the absorbance of *E. coli* suspensions in LB broth containing 0.01 to 1 μM cat-MF was recorded over 4 h at 37 °C. As discussed in Sect. 2.2.2, increased absorbance in a culture is a good indicator for bacterial growth. Given that *E. coli* has a generation time of 20 min at 37 °C [24], monitoring the absorbance over 4 h should yield reliable results regarding bacterial growth. After 4 h, suspensions of *E. coli* grown in the presence of 0.5 and 1 μM had an absorbance of 0.27 compared to an absorbance of approximately 0.35 in the untreated control and *E. coli* treated with lower concentrations of cat-MF (Fig. 6.9). Therefore, bacterial growth was only mildly affected by elevated cat-MF exposure after 4 h.

It is possible that longer incubation periods or higher cat-MF concentrations would result in a more pronounced effect on growth inhibition. Previously, incubation of *S. aureus* with 3 mg mL^{-1} SPIONs (100 nm diameter) for 12 h resulted in a decrease in the number of live bacteria [25], and carboxyl-functionalised SPIONs (25 nm diameter) have been shown to penetrate the outer membrane of *E. coli* and disrupt its structural integrity after a 12 h exposure [9]. Similarly, exposure to silica-coated iron oxide nanoparticles resulted in growth inhibition of both *E. coli* and *S. aureus* during a 6 h incubation period [26]. Furthermore, the small size of cat-MF (12 nm) may also facilitate penetration of the outer membrane of *E. coli*. For example, 10 to 80 nm iron nanoparticles have been shown to penetrate the outer membrane of *E. coli* leading to growth inhibition [27].

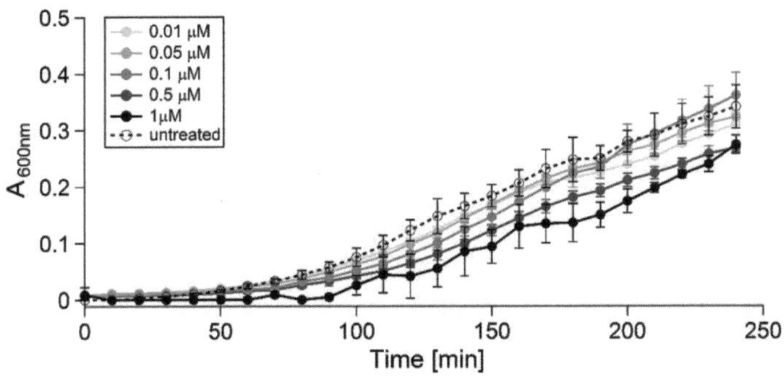

Fig. 6.9 Growth curve of *E. coli* incubated with a range of cat-MF concentrations. Absorbance at 600 nm was slightly reduced in *E. coli* grown in the presence of 0.5 and 1 μM cat-MF compared to untreated cells. Average and standard deviation from two replicates are shown

6.5 Conclusions and Future Directions

It was demonstrated that cat-MF can be used to magnetise *E. coli* within a 1 min incubation period, such that 99.9% of the magnetically labelled bacteria were removed using MACS. It is the first time that MACS columns have been used to capture magnetically labelled bacteria, and it was found that this method resulted in higher capture efficiency compared to conventional setups. Growth of *E. coli* in the presence of 0.5 and 1 μM cat-MF was mildly reduced after a 4 h incubation period. It remains to be investigated if higher incubation concentrations or longer exposure times could result in more significant growth inhibition, and whether this effect is mediated by cationisation of MF or the metal oxide core of the nanoparticle.

Although capture efficiency of the MACS columns was over 99.9%, complete capture of *E. coli* was only achieved when samples contained low numbers of bacteria (10^2 cfu mL^{-1} and less). Increasing the labelling time or the concentration of cat-MF did not improve capture efficiency. Interestingly, immobilisation of cat-MF labelled *S. aureus* was found to be more efficient, and it remains to be clarified whether this was due to differences in size or cellular motility. Further experiments using non-flagellated strains of *E. coli* could help elucidate this point. In addition, it could be explored whether a "second pass" would improve capture efficiency, *i.e.* passing the flow-through over the column a second time, or whether using a longer column would result in complete capture.

References

1. Morens, D.M., Folkers, G.K., Fauci, A.S.: The challenge of emerging and re-emerging infectious diseases. Nature **430**(6996), 242–249 (2004)
2. Naravaneni, R., Jamil, K.: Rapid detection of food-borne pathogens by using molecular techniques. J. Med. Microbiol. **54**(1), 51–54 (2005)
3. Boehm, D.A., Gottlieb, P.A., Hua, S.Z.: On-chip microfluidic biosensor for bacterial detection and identification. Sens. Actuators B: Chem. **126**(2), 508–514 (2007)
4. Peytavi, R., et al.: Microfluidic device for rapid (<15 Min) automated microarray hybridization. Clin. Chem. **51**(10), 1836–1844 (2005)
5. Kell, A.J., et al.: Vancomycin-modified nanoparticles for efficient targeting and preconcentration of gram-positive and gram-negative bacteria. ACS Nano **2**(9), 1777–1788 (2008)
6. Edberg, S., Rice, E., Karlin, R., Allen, M.: Escherichia coli: the best biological drinking water indicator for public health protection. J. Appl. Microbiol. **88**(S1), 106S–116S (2000)
7. Dickson, J.S., Koohmaraie, M.: Cell surface charge characteristics and their relationship to bacterial attachment to meat surfaces. Appl. Environ. Microbiol. **55**(4), 832–836 (1989)
8. Huang, Y.-F., Wang, Y.-F., Yan, X.-P.: Amine-functionalized magnetic nanoparticles for rapid capture and removal of bacterial pathogens. Environ. Sci. Technol. **44**(20), 7908–7913 (2010)
9. Singh, S., Barick, K., Bahadur, D.: Surface engineered magnetic nanoparticles for removal of toxic metal ions and bacterial pathogens. J. Hazard. Mater. **192**(3), 1539–1547 (2011)
10. El-Boubbou, K., Gruden, C., Huang, X.: Magnetic glyco-nanoparticles: a unique tool for rapid pathogen detection, decontamination, and strain differentiation. J. Am. Chem. Soc. **129** (44), 13392–13393 (2007)

11. Organization, W.H.: Guidelines for Drinking-Water Quality: Recommendations, vol. 1. World Health Organization (2004)
12. Carreira, S.C., Spencer, J., Schwarzacher, W., Seddon, A.: Cationized magnetoferritin enables rapid labeling and concentration of gram-positive and gram-negative bacteria in magnetic cell separation columns. Appl. Environ. Microbiol. **82**(12), 3599–3604 (2016)
13. Anderson, K.L.: Cationized ferritin as a stain for electron microscopic observation of bacterial ultrastructure. Biotech. Histochem. **73**(5), 278–288 (1998)
14. Weiss, R., Schiefer, H.-G., Krauss, H.: Ultrastructural visualization of Klebsiella capsules by polycationic ferritin. FEMS Microbiol. Lett. **6**, 435–437 (1979)
15. Hammond, S.M., Lambert, P., Rycroft, A.: The Bacterial Cell Surface. Springer Science & Business Media (2012)
16. Schiefer, H.-G., Krauss, H., Brunner, H., Gerhardt, U.: Ultrastructural visualization of anionic sites on mycoplasma membranes by polycationic ferritin. J. Bacteriol. **127**(1), 461–468 (1976)
17. Wagner, S., Robinette, D.: Evaluation of an automated microbiologic blood culture device for detection of bacteria in platelet components. Transfusion **38**(7), 674–679 (1998)
18. Gu, H.W., Xu, K.M., Xu, C.J., Xu, B.: Biofunctional magnetic nanoparticles for protein separation and pathogen detection. Chem. Commun. **9**, 941–949 (2006)
19. Silhavy, T.J., Kahne, D., Walker, S.: The bacterial cell envelope. Cold Spring Harbor Perspect. Biol. **2**(5), a000414 (2010)
20. Mitik-Dineva, N., et al.: Escherichia Coli, Pseudomonas aeruginosa, and Staphylococcus aureus attachment patterns on glass surfaces with nanoscale roughness. Curr. Microbiol. **58** (3), 268–273 (2009)
21. Gu, H., Ho, P.-L., Tsang, K.W., Wang, L., Xu, B.: Using biofunctional magnetic nanoparticles to capture vancomycin-resistant Enterococci and other gram-positive bacteria at ultralow concentration. J. Am. Chem. Soc. **125**(51), 15702–15703 (2003)
22. Winn, W.C., Koneman, E.W.: Koneman's Color Atlas and Textbook of Diagnostic Microbiology. Lippincott Williams & Wilkins (2006)
23. Otto, C.C., Cunningham, T.M., Hansen, M.R., Haydel, S.E.: Effects of antibacterial mineral leachates on the cellular ultrastructure, morphology, and membrane integrity of Escherichia coli and methicillin-resistant Staphylococcus aureus. Ann. Clin. Microbiol. Antimicrob. **9**(1), 26 (2010)
24. Doyle, M., Schoeni, J.: Survival and growth characteristics of Escherichia coli associated with hemorrhagic colitis. Appl. Environ. Microbiol. **48**(4), 855–856 (1984)
25. Tran, N., Mir, A., Mallik, D., Sinha, A., Nayar, S., Webster, T.J.: Bactericidal effect of iron oxide nanoparticles on Staphylococcus aureus. Int. J. Nanomed. **5**, 277 (2010)
26. Darwish, M.S., Nguyen, N.H., Ševců, A., Stibor, I.: Functionalized magnetic nanoparticles and their effect on Escherichia coli and Staphylococcus aureus. J. Nanomater. (2015)
27. Lee, C., Kim, J.Y., Lee, W.I., Nelson, K.L., Yoon, J., Sedlak, D.L.: Bactericidal effect of zero-valent iron nanoparticles on Escherichia coli. Environ. Sci. Technol. **42**(13), 4927–4933 (2008)

Chapter 7
Overall Conclusions and Future Opportunities

The primary aim of this work was to develop a facile surface functionalisation for magnetoferritin (MF) that would enable rapid and versatile magnetic cell labelling. Previously reported labelling protocols require long incubation periods and sometimes laborious surface functionalisations. Here, it was shown for the first time that efficient magnetic labelling could be achieved within a one minute incubation period using low SPION concentrations ($\sim 10\ \mu g\ mL^{-1}$ iron). This was accomplished by cationising acidic amino acid residues on the MF surface, which generated a novel, cationic SPION that readily adsorbed to anionic domains on the cell surface, such as proteoglycans. This non-specific labelling mechanism was successful in rapidly and efficiently magnetising hMSCs, as well as gram-positive and gram-negative bacteria, demonstrating the versatility of this approach.

The magnetised hMSCs exhibited lasting MRI contrast, and exposure to cationised MF (cat-MF) did not adversely affect cell viability, membrane integrity or differentiation capacity. Therefore, this novel SPION system represents an attractive candidate for tracking stem cell therapy using MRI without impairing the regenerative capacity of the labelled cells. Significantly, chondrogenesis was not inhibited, which is usually affected by SPION exposure. The underlying reason for this result is an interesting subject for future study. One hypothesis is that the administration of SPIONs in an iron storage protein affects intracellular processing, possibly leading to a slower release of iron ions over time. In this context, the fate of the protein cage after internalisation is also unclear and remains to be elucidated.

The non-specific labelling mechanism should enable cat-MF to rapidly magnetise a range of different cell types. Indeed, a 15 min incubation with cat-MF achieved magnetisation of both *E. coli* and *S. aureus*, such that these bacteria could be immobilised in MACS columns with an efficiency of up to 99.9999%. These capture efficiencies are unprecedented and are likely due to the use of the MACS column setup, which has not yet been explored in the context of bacterial capture. Furthermore, this method was used to concentrate labelled *E. coli* from a dilute cell suspension ($<10^2$ cfu mL^{-1}), demonstrating its suitability as a pre-processing step for rapid, microfluidics-based pathogen detection.

S. Correia Carreira, *Rapid Cell Magnetisation Using Cationised Magnetoferritin*, Springer Theses, DOI 10.1007/978-3-319-60333-9_7

It should be noted, however, that the upscaling of MF synthesis and functionalisation may be more costly than conventional SPION synthesis protocols, because protein preparation and purification methods require lengthy procedures and a variety of specialised pieces of equipment. Furthermore, approval for clinical use may also be hard to obtain due to the protein's animal origin. Thus, ApoF samples are subject to individual variation, and during the course of this study variations in mineralisation efficiencies were indeed observed. This represents a problem, because in order to obtain approval for clinical use, a high degree of reproducibility needs to be demonstrated. However, with the rapid progress made in synthetic biology, it may be possible to fabricate artificial ApoF cages that yield reproducibly high mineralisation efficiencies.

Finally, ApoF is a highly flexible vector that can be loaded with alternative functional molecules, thereby extending the presented concepts to applications beyond magnetic labelling. For example, radioisotopes [1], quantum dots [2] and anti-cancer drugs [3, 4] have been loaded into ApoF. Cationisation could enhance the delivery efficiency of these cargoes, for example when they are directly injected into the tissue. Furthermore, not only the ApoF cavity, but also the protein's surface could be loaded with other functional molecules. For example, bioactive molecules could be linked to the MF surface, such that spatial gradients of these "chemical signals" could be created using an external magnetic field. This would be of great interest in tissue engineering, because spatially patterned signals can influence cell migration and alignment. For example, it has been shown that nerve regeneration benefitted from spatial gradients of nerve growth factor created in a hydrogel scaffold [5]. In a preliminary experiment, a spatial gradient was created by immersing a collagen scaffold in a solution of fluorescein-conjugated MF and applying an external magnetic field (generated by a permanent magnet). The fluorophore acted as a model signal molecule and the gradient could be visualised using confocal microscopy. Image analysis revealed that a spatial gradient could indeed be created, however, it was not very pronounced (Fig. 7.1). Applying stronger magnetic fields could improve gradient formation. Ultimately, using such a "magnetic patterning" approach would enable the customisation of commercially available scaffolds with chemical gradients.

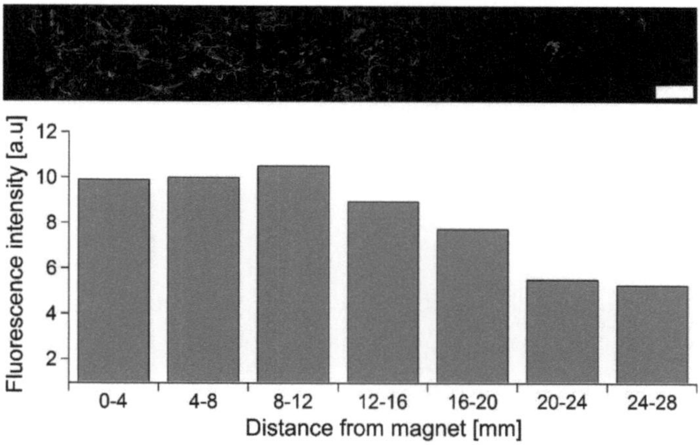

Fig. 7.1 Creating spatial gradients with MF. Carboxy-fluorescein was covalently linked *via* a peptide bond to amine groups on the MF surface. A collagen scaffold was immersed in a solution of fluorescent MF and a permanent magnet (~0.7 T) was placed against the dish over night. Confocal microscopy and image analysis revealed that a spatial gradient could be generated

References

1. Lin, X., et al.: Chimeric ferritin nanocages for multiple function loading and multimodal imaging. Nano Lett. **11**(2), 814–819 (2011)
2. Wong, K.K., Mann, S.: Biomimetic synthesis of cadmium sulfide-ferritin nanocomposites. Adv. Mater. **8**(11), 928–932 (1996)
3. Xing, R.M., et al.: Characterization and cellular uptake of platinum anticancer drugs encapsulated in apoferritin. J. Inorg. Biochem. **103**(7), 1039–1044 (2009)
4. Liang, M., et al.: H-Ferritin–Nanocaged doxorubicin nanoparticles specifically target and kill tumors with a single-dose injection. Proc. Natl. Acad. Sci. **111**(41), 14900–14905 (2014)
5. Dodla, M.C., Bellamkonda, R.V.: Differences between the effect of anisotropic and isotropic laminin and nerve growth factor presenting scaffolds on nerve regeneration across long peripheral nerve gaps. Biomaterials **29**(1), 33–46 (2008)

Appendix A
List of Naturally Occurring Amino Acids

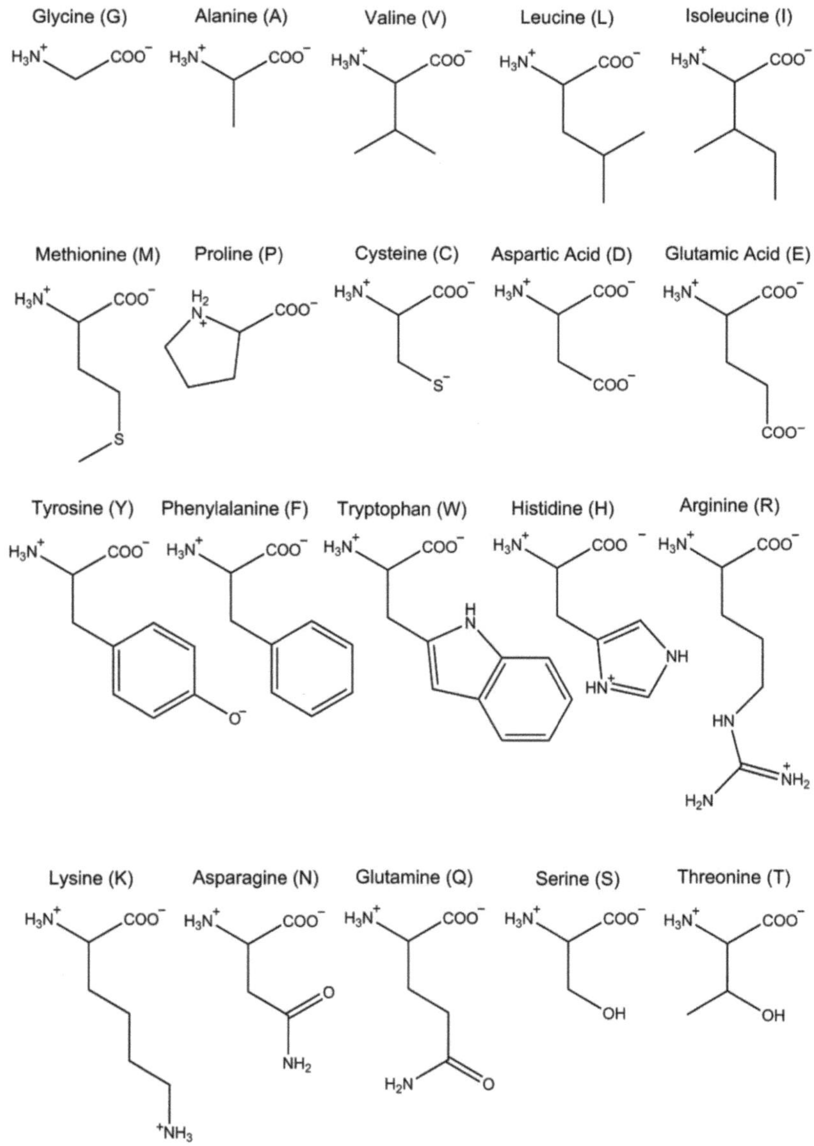

S. Correia Carreira, *Rapid Cell Magnetisation Using Cationised Magnetoferritin*,
Springer Theses, DOI 10.1007/978-3-319-60333-9

Appendix B
Background Subtraction with the MPMS

To measure the magnetic moment of MF and cat-MF with the MPMS, a voltage curve is recorded as the sample is passed through the detection coils. The ideal voltage response curve of a point-source magnetic dipole moving through a second-order gradiometer (i.e. the configuration in which the detection coils are set up) is shown in Fig. B.1.

However, this ideal signal is only observed when the sample is much smaller than the detection coils and if it is uniformly magnetised. Non-uniform magnetisation and the size and shape of the sample can lead to deviation from this ideal shape. Furthermore, when the sample signal is comparable to the signal produced by the sample holder, the response function will not have the ideal shape either. This can be seen in Fig. B.2, in which the voltage curve for an MF sample and the voltage curve for the sample holder alone are shown for an applied field of 0.5 T.

To subtract the sample holder background, the sample and the empty sample holder were measured separately but under identical conditions (i.e. the sample holder was mounted in the exact same location for both measurements). The data sets of the two voltage curves were then subtracted point by point to yield a

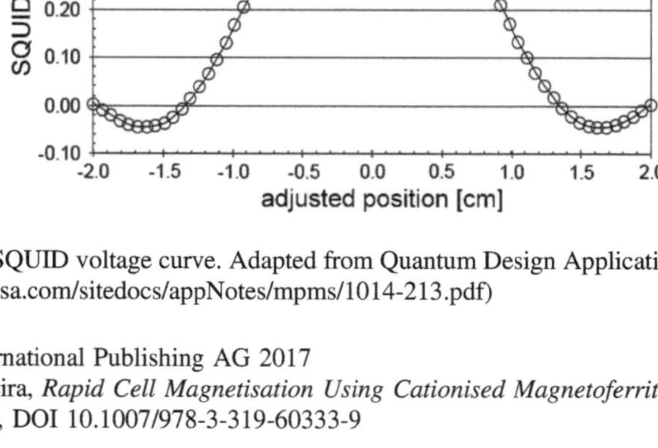

Fig. B.1 Ideal SQUID voltage curve. Adapted from Quantum Design Application Note 1014-213 (http://www.qdusa.com/sitedocs/appNotes/mpms/1014-213.pdf)

S. Correia Carreira, *Rapid Cell Magnetisation Using Cationised Magnetoferritin*, Springer Theses, DOI 10.1007/978-3-319-60333-9

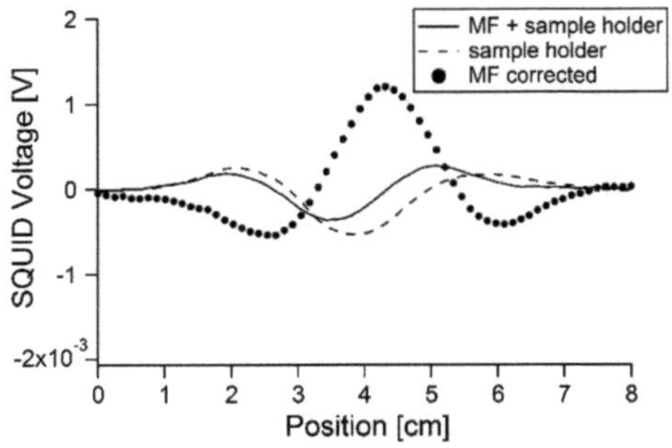

Fig. B.2 Representative SQUID voltage curves recorded at 0.5 T. The contribution of the sample holder alone to the total voltage curve ("MF+sample holder") was measured separately for all applied magnetic field strength and subsequently subtracted from the total signal to yield the corrected MF voltage curve.

background-corrected voltage curve (see Fig. B.2). Thus, the background subtraction procedure removed the contribution of the sample holder and produced a response, to which the following equation was fitted:

$$f(Z) = X_1 + X_2 Z + X_3\{2\left[R^2 + (Z + X_4)^2\right]^{-3/2} - [R^2 + (\Lambda + (Z + X_4)^2]^{-3/2} - [R^2 + (-\Lambda + (Z + X_4)^2]^{-3/2}\}$$

where *f(Z)* is the SQUID voltage as a function of the sample position *Z*. The constants in this equation are the longitudinal radius, *R* (0.97 cm), and the longitudinal coil separation, Λ (1.519 cm), of the detection coils. The fit parameter X_1 is a constant offset voltage, X_2 takes into account the linear electronic SQUID drift during data collection, and X_4 is the shift of the sample along the magnet axis during the measurement. The magnetic moment, *M*, was calculated from the amplitude X3 by applying the following equation:

$$M[emu] = X_3/0.9125$$

The value 0.9125 is a constant correction factor supplied by the manufacturer. The resulting unit for the magnetic moment was given as the cgs unit *emu*. This was converted to the SI unit of the magnetic moment (Am^2) using the appropriate conversion factor: 1 Am^2 = 10^3 emu.

Appendix C
Adipo- and Osteogenesis of Untreated hMSCs

Untreated hMSCs differentiated into adipocytes and osteoblasts upon culture in differentiation medium (Fig. C.1a, c), but remained undifferentiated when cultured in regular medium (Fig. C.1b, d).

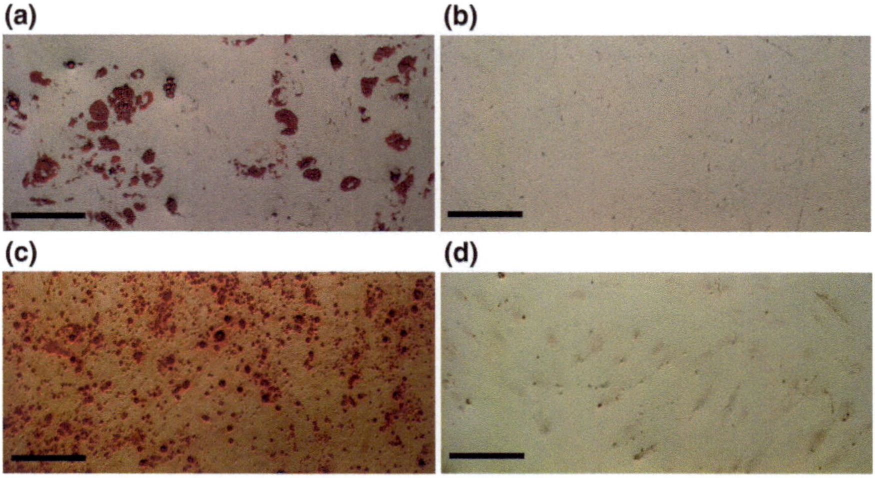

Fig. C.1 Monolayer differentiation of untreated hMSCs. Representative bright field microscopy images of **a** hMSC-derived adipocytes with fatty acid vacuoles stained using Oil Red. **b** Undifferentiated hMSCs stained with Oil Red. No fatty vacuoles are visible. **c** hMSC-derived osteoblasts with visible calcium phosphate deposits stained using Alizarin Red. **d** Undifferentiated hMSCs stained with Alizarin Red; no calcium phosphate deposits are visible. All scale bars: 100 μm

S. Correia Carreira, *Rapid Cell Magnetisation Using Cationised Magnetoferritin*, Springer Theses, DOI 10.1007/978-3-319-60333-9

Appendix D
Unstained and H&E Stained Cartilage Sections

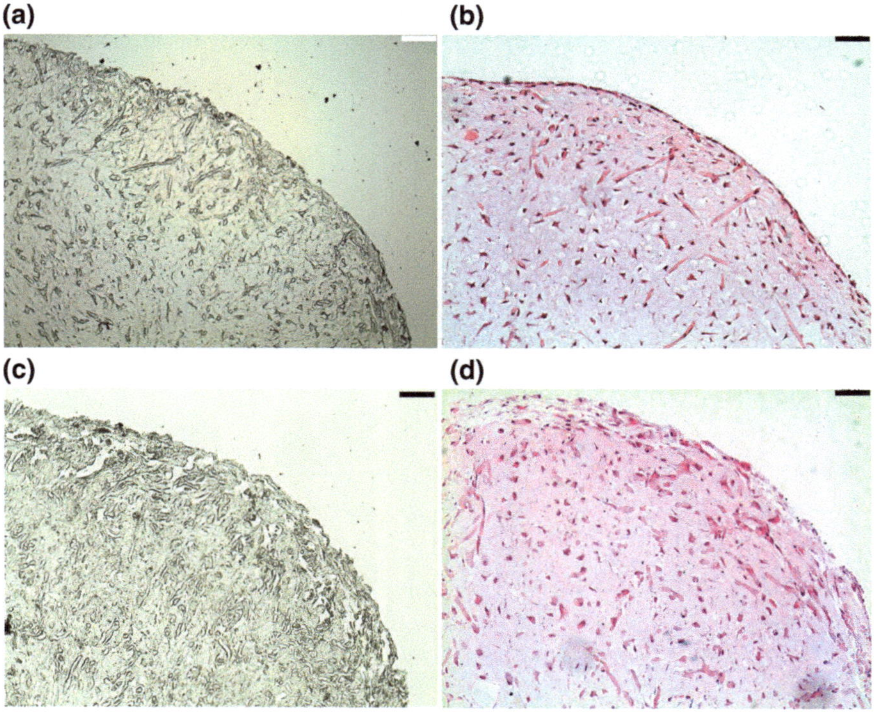

FIg. D.1 Histological sections of engineered cartilage. Representative bright field microscopy images of unstained sections of **a** cartilage engineered from untreated hMSCs and **c** engineered from hMSCs exposed to 0.5 μM cat-MF for 30 min. H&E stained sections of cartilage grown from **b** untreated hMSC show that cells were evenly distributed throughout the tissue. **d** The same was the case in cartilage grown from cat-MF-treated hMSCs. All *scale bars* 100 μm

Unstained sections of cartilage constructs engineered from untreated and cat-MF-treated hMSCs are shown in Fig. D.1a, c. H&E stain revealed an even distribution of cells throughout the cartilage construct in both treatment groups (Fig. D.1b, d).

S. Correia Carreira, *Rapid Cell Magnetisation Using Cationised Magnetoferritin*, Springer Theses, DOI 10.1007/978-3-319-60333-9

Curriculum Vitae

Education

PhD in Functional Nanomaterials Sep 2011–Aug 2015
University of Bristol, Centre for Functional Nanomaterials, Bristol, UK

Diplom in Biology Oct 2001–March 2007
Humboldt University, Berlin, Germany

Research Experience

School of Oral and Dental Sciences, University of Bristol, UK since Oct 2016
Postdoctoral Research Associate, supervised by Dr. Michele Barbour
Developing topical formulations of chlorhexidine hexametaphosphate particles, and testing their stability and antimicrobial efficacy after mechanical challenge on artificial skin substrates

H.H. Wills Physics Laboratory, University of Bristol, UK May 2016–Aug 2016
Postdoctoral Research Associate, supervised by Prof. Walther Schwarzacher
Building and testing an electrochemical quartz crystal microbalance setup to measure bacterial adsorption onto charged surfaces

S. Correia Carreira, *Rapid Cell Magnetisation Using Cationised Magnetoferritin*, Springer Theses, DOI 10.1007/978-3-319-60333-9

Merchant Venturers' School of Engineering, University of Bristol, UK Sep 2015–Mar 2016

Postdoctoral Research Assistant, supervised by Dr. Sabine Hauert

Engineering of a tumour-on-a-chip model to study nanoparticle dynamics in brain tumour tissue

Bristol Centre for Functional Nanomaterials, University of Bristol, UK Sep 2011–Aug 2015

Postgraduate Researcher, supervised by Dr. Annela Seddon

Development of protein-based magnetic nanoparticles for rapid magnetisation of stem cells and bacteria, including tissue engineering with magnetised stem cells

University Hospitals Bristol NHS Foundation Trust, Bristol, UK Mar 2008–Aug 2011

Healthcare Scientist, supervised by Dr. Margaret Saunders

Research into placental transport and toxicity of polychlorinated biphenyls (PCBs) and nanomaterials and evaluation of *in vitro* toxicity testing strategies for nanomaterials

Leibniz Institute for Zoo and Wildlife Research, Berlin, Germany Sep 2005–Mar 2007

Diplom Research Student, supervised by Dr. Christian Voigt

Validation of a minimally invasive blood-sampling technique for use in small mammals using blood-sucking bugs

Grants and awards

Faculty of Science Doctoral Prize for Physical Sciences, University of Bristol **Sep 2016**

Awarded for the best thesis submitted in 2015/16

EPSRC Bristol Bridge research grant (£ 14,000), University of Bristol **May 2016**

Research grant to undertake proof-of-principle project investigating bacterial adsorption onto charged surfaces

Prize for best talk, INASCON (International Nanoscience Student Conference), Denmark **Aug 2014**

George Pugh Memorial Prize, University of Bristol **Jul 2014**

Awarded for outstanding contribution to the life of the School of Physics and the wider student community

Prize for best poster, China NANO conference, China **Sep 2013**

Leonardo da Vinci Postgraduate Mobility grant (2000 €) Nov 2007
Awarded from the European Union to undertake an internship at the Knowledge Transfer Office of the Centro de Investigación Principe Felipe, Valencia, Spain

Publications

Peer-Reviewed Articles

Correia Carreira, S., Spencer, J., Schwarzacher, W., Seddon, A.M.: Cationized Magnetoferritin Enables Rapid Labeling and Concentration of Gram-Positive and Gram-Negative Bacteria in Magnetic Cell Separation Columns. Appl. Environ. Microbiol **82**(12), 3599–3604 (2016). doi:10.1128/AEM.00720-16

Correia Carreira, S., Armstrong, J.P.K., Seddon, A.M., Perriman, A.W., Hartley-Davies, R., Schwarzacher, W.: Ultra-fast stem cell labelling using cationised magnetoferritin. Nanoscale **8**, 7474–7483 (2016). doi:10.1039/C5NR07144E

Correia Carreira, S., Armstrong, J.P.K., Okuda, M., Seddon, A.M., Perriman, A. W., Schwarzacher, W.: Synthesis of cationised magnetoferritin for ultra-fast magnetization of cells. J Vis. Exp **118** (2016). doi:10.3791/54785

Chagas, E.F., **Correia Carreira, S.**, Schwarzacher, W.: Using magnetic nanoparticles to probe protein damage in ferritin caused by freeze concentration. AIP Adv **5**, 117201. doi:10.1063/1.4935261

Cowie, H., Magdolenova, Z., Saunders, M., Drlickova, M., **Correia Carreira, S.**, Halamoda Kenzaoi, B., et al.: Suitability of human and mammalian cells of different origin for the assessment of genotoxicity of metal and polymeric engineered nanoparticles. Nanotoxicology **9**(S1), 57–65 (2015). doi:10.3109/17435390.2014.940407

Dusinska, M., Boland, S., Saunders, M., Juillerat-Jeanneret, L., Tran, L., Pojana, G., Marcomini, A., Volkovova, K., Tulinska, J., Knudsen, LE., Gombau, L., Whelan, M., Collins, AR., Marano, F., Housiadas, C., Bilanicova, D., Halamoda Kenzaoui, B., **Correia Carreira, S.**, et al.: Towards an alternative testing strategy for nanomaterials used in nanomedicine: lessons from NanoTEST. Nanotoxicology **9**(S1), 118–132 (2015). doi:10.3109/17435390.2014.991431

Correia Carreira, S., Walker, L., Paul, K., Saunders, M.: The toxicity, transport and uptake of nanoparticles in the in vitro BeWo b30 placental cell barrier model used within NanoTEST. Nanotoxicology **9**(S1), 66–78 (2015). doi:10.3109/17435390.2013.833317

Eloi, J.C., Okuda, M., **Correia Carreira, S.**, Schwarzacher, W., Correia, M.J., Figueiredo, W.: Effective energy barrier distributions for random and aligned magnetic nanoparticles. J. Phys.: Condens. Matter **26**(14), 146006 (2014). doi:10.1088/0953-8984/26/14/146006

Saunders, M., Magnanti, B.L., **Correia Carreira, S.**, Yang, A., Alamo-Hernandez, U., Riojas-Rodriguez, H., et al.: Chlorpyrifos and neurodevelopmental effects: a literature review and expert elicitation on research and policy. Environ. Health **11** (1), S5 (2012). doi:10.1186/1476-069X-11-S1-S5

Correia Carreira, S., Cartwright, L., Mathiesen, L., Knudsen, L.E., Saunders, M.: Studying placental transfer of highly purified non-dioxin-like PCBs in two models of the placental barrier. Placenta **32**(3), 283–291 (2011). doi:10.1016/j.placenta.2010.12.024

Book Chapters

Saunders, M., Hutchinson, G., **Correia Carreira, S.**: Reproductive toxicity of nanomaterials. In: Handbook of Safety Assessment of Nanomaterials. From Toxicological Testing to Personalized Medicine, Chap. 16, pp. 461–498. Pan Stanford Publishing (2015)

Dusinska, M., Runden-Pran, E., **Correia Carreira, S.**, Saunders, M.: Critical evaluation of toxicity tests. In: Adverse Effects of Engineered Nanomaterials, Chap. 2, pp. 63–83. Elsevier Academic Press (2012)

MIX
Papier aus verantwortungsvollen Quellen
Paper from responsible sources
FSC® C105338

If you have any concerns about our products,
you can contact us on
ProductSafety@springernature.com

In case Publisher is established outside the EU,
the EU authorized representative is:
Springer Nature Customer Service Center GmbH
Europaplatz 3, 69115 Heidelberg, Germany

Printed by Libri Plureos GmbH
in Hamburg, Germany